CROSS-BORDER COOPERATION STRUCTURES IN EUROPE

LEARNING FROM THE PAST, LOOKING TO THE FUTURE

P.I.E. Peter Lang
Bruxelles · Bern · Berlin · Frankfurt am Main · New York · Oxford · Wien

EUROCLIO is a scientific and editorial project, a network of research institutions and researchers, and an ideas forum. EUROCLIO as an editorial project consists of two aspects: the first concerns studies and documents, the second concerns tools. Both are aimed at making the results of research more accessible, and also at opening up paths through the history of European construction/integration/unification.

The EUROCLIO series meets a dual objective:

- to provide reference tools for research,
- to provide a platform for this research in terms of the publication of results.

The series thus consists of two sub-series that satisfy these requirements: the Studies and Documents series and the References series. These two series are aimed at general libraries and/or university history departments, teachers and researchers, and in certain cases, specific professional circles.

The Studies and Documents series consists of monographs, collections of articles, conference proceedings, and collections of texts with notes for teaching purposes.

The References series consists of bibliographies, guides and other tools. It thus contributes to the creation of a database making up a "Permanent catalogue of sources and bibliographies on European construction".

CROSS-BORDER COOPERATION STRUCTURES IN EUROPE

LEARNING FROM THE PAST, LOOKING TO THE FUTURE

Luis DOMÍNGUEZ & Iva PIRES (eds.)

Euroclio
No. 82

This project has been funded with support from the European Commission. This publication reflects the views only of the authors, and the Commission cannot be held responsible for any use which may be made of the information contained therein.

This publication has been peer-reviewed.

Editions scientifiques internationales
Brussels, 2014
1 avenue Maurice, B-1050 Brussels, Belgium
pie@peterlang.com ; www.peterlang.com

ISSN 0944-2294
ISBN 978-2-87574-186-8 (paperback)
ISBN 978-3-0352-6452-4 (eBook)
D/2014/5678/56

CIP available from the British Library, UK and from the Library of Congress, USA.

"Die Deutsche National Bibliothek" lists this publication in the "Deutsche Nationalbibliografie"; detailed bibliographic data is available on the Internet at http://dnb.de.

Table of Contents

Introduction

Cross-border studies and their political, economic, social, and cultural relevance have a significant presence in the scientific literature, although not with the same intensity among the different social sciences, nor among countries. In fact, the Mexico-US border has had, for a long time, an iconic status within border studies; EU borders were later integrated in cross-border research and, more recently, borders in other continents. At the same time, an increasing number of collective studies characterised by the interdisciplinarity of their approach have emerged in the last decades. Along this path, the interrelation of several social sciences involved in border studies has enabled the achievement of significant conceptual and methodological contributions. Thus, anthropologists managed to include border citizens as active participants in their studies[1], breaking away from the classical state and institution centred approaches. Geographers introduced the concept of border landscapes in order to define the space located in the milieu of each state's limes, shaped by the constant interaction between the people who live there and by the geographical features of the border area[2]. Along these lines, historians coined the notion of borderland in order to highlight the relevance of these territories and their people in building nation-states and their interrelationships, from the state towards its borders and vice versa[3], in a dual relationship in which the political decisions taken by national actors predetermine border life and context, but, at the same time, state decision-making is also determined by the border itself and the actors involved in it. In addition to this, political scientists have refined the notion of borderland as a unique space for developing processes of ethnic, local, class, religious and linguistic identity and their different meanings[4],

---

1 The study of the Italian Tyrol, with its successive historical partitions and the subsequent transformation of local political loyalties, constituted an innovation in this respect, J. Colea, and E. R. Wolf, *The Hidden Frontier: Ecology and Ethnicity in an Alpine Valley*, New York, Academic Press, 1974.

2 J. R. V. Prescott, *Political Frontiers and Boundaries*, London, Unwin Hyman, 1987. Previously, J. R. V. Prescott, *Boundaries and Frontiers*, London, Croom Helm, 1978.

3 O. J. Martínez, *Border People: Life and Society in the US-Mexico Borderlands*, Tucson, University of Arizona Press, 1994; M. Baud, and W. van Schendel, "Toward Comparative History of Borderlands", in *Journal of World History*, Vol. 8, No. 2, 1997, pp. 211-242.

4 M. Anderson, *Frontiers: Territory and State Formation in the Modern World*, Cambridge, Polity, 1996; M. Perkmann, and N. L. Sum (eds.), *Globalization, Regionalization and Cross-Border Regions*, London, Macmillan, 2002.

and have contributed to the acknowledgement of culture as a key facilitator for cross-border cooperation strategies. For sociologists, focused on the study of social groups and institutions, the main finding might be that of ambivalent identities, which are characteristic of border societies open to influences increased by various formal and informal cooperation processes[5], this identity ambiguity of border residents undermines the classic concept of sovereignty, based on the identification of state with territory[6].

In the case of Europe, border studies have been characterized by a special feature from the outset of the post-war integration process, the emergence of a range of formal and informal cooperation structures with different degrees of institutionalisation, either by means of grassroots initiatives (bottom up), or through supranational and intergovernmental initiatives (top down), with uneven results in its management. These structures have emerged in different historical moments and contexts as a response to diverse motivations. Initially, cultural affinities and a common desire for reconciliation allowed for early signs of cross-border cooperation. In the 1960s, the so much in vogue spatial planning enhanced cooperation as the appropriate formula to promote regional development, also on borders. The International Congress of Regional Planners in 1965 was an important milestone; Regio Basiliensis summoned a thousand experts and professionals (political authorities, scholars and spatial and urban planning analysts) with the presence of other ground-breaking structures of cross-border cooperation, for example Euregio, CIMAB and the first associations of what later became the Regio Rhein-Waal[7]. The 1970s are known as a period of decentralization[8] caused by the double crises of the Keynesian and Fordist models[9]. Although it is true that the regionalisation process in Europe did not lead to the emergence of levels of regional government with elected representatives in most of the countries, it did allow for the emergence of regional power and a considerable increase in the networks of cooperation structures in the following decade; at this point, interest in cooperation had a strong

5 R. Strassoldo, "Boundaries in Sociological Theory: a Reassessment", in R. Strassoldo and G. Delli Zotti (eds.), *Cooperation and Conflict in Border Areas*, Milan, Franco Angeli, 1982.

6 L. O'Dowd, "From a "Borderless World" to a "World of Borders": "Bringing History back in", in *Environment and Planning D: Society and Space*, Vol. 28, 2010, pp. 1031-1050.

7 It is worth mentioning that in the previous year, the Conference of Local Authorities of the Council of Europe had adopted Resolution 46 (1964), with regard to spatial planning in border regions.

8 Y. Meny (dir.), *Dix ans de régionalisation en Europe. Bilan et perspectives*, Paris, Cujas, 1982.

9 C. Trigilia, "Il paradosso delle regione? Regolazione economica e rappresentanza degli interesi", in *Meridiana*, Vol. 6, 1989.

functional component, it was about joining efforts towards common objectives, such as economic development and promotion, infrastructure, environment and culture. As a result of the European integration process, peripheral and border territories were given the opportunity to become Euroregions, which acquired centrality in the new European space in the making[10].

A. Paasi[11] has pointed out that the 1990s witnessed a remarkable interest in borders with an interdisciplinary approach from different academic fields. Certainly, there are several factors which can explain this interest, including the implosion of the Soviet Empire, the reconfiguration of European borders and the implementation of cohesion policies on the part of the EU, with large allocations of structural funds via initiatives such as INTERREG, and to a lesser extent, PHARE and TACIS, specifically designed for cross-border cooperation. In Europe, the study of regionalism has run in parallel with border studies. In fact, the concept of neo-regionalism, closely linked to the process of economic globalization, was coined in this decade. Neo-regionalism differs from classic regionalism – related to regulatory (federalism) or political (nationalism) actions – in that an emphasis is placed on public policies and their efficiency and, especially, in that it substitutes intergovernmental relations for networks of collective action, which associate private actors and public authorities from different levels within border-based cooperation processes[12]. In this context, the concept of paradiplomacy[13] emerged in order to describe regions' external actions. Cross-border cooperation in the EU has become a major field of research, and it is much more developed than the one in North America, this is due to the fact that the EU and its cooperation programmes offer resources which reduce cooperation costs, provide political support against national governments and facilitate management and assistance in organising initiatives[14]. However, the booming of studies, theories and concepts has not always left a lasting trace, one of the best

10 R. Balme, "Pourquoi le gouvernement change-t-il d'échelle?", in R. Balme (dir.), *Les politiques du néo-régionalisme*, Paris, Economica, 1996, p. 25 [11-39]. This collective work contains important contributions about cooperation structures, which focus on transnational rather than border cooperation.

11 A. Paasi, "Inclusion, Exclusion and Territorial Identities: the Meanings of Boundaries in the Globalizing Geopolitical Landscape", in *Nordisk Samhällsgeografisk Tidskrift*, Vol. 23, 1996, pp. 5, 6 [3-17].

12 R. Balme, "Pourquoi le gouvernement change-t-il d'échelle?", in R. Balme (dir.), *Les politiques du néo-régionalisme*, Paris, Economica, 1996, p. 35.

13 F. Aldecoa and M. Keating, *Paradiplomacy in Action. The Foreign Relations of Subnational Governments*, London, Frank Cass, 1999.

14 M. Keating, "Regions and International Affairs: Motives, Opportunities and Strategies", in *Regional and Federal Studies*, Vol. 9, 1999, p. 10 [1-16].

examples might be the idea of the de-territorialisation of politics[15], swept aside in favour of re-territorialisation[16].

The first years of this century have been very productive in terms of collective works about regions and border studies; however, less attention has been given to cross-border cooperation, although there are significant contributions[17]. In Europe, the study of cross-border cooperation counts on important contributions from a scientific field often absent from other geographical areas, i.e. the legal field, concerned with the complexity of the newly created structures. *Le Réseau d'étude des normes transfrontalières et inter-territoriales* (RENTI), established in Bayonne in 2004, illustrates this effort through its significant contributions[18]. Innovation has also occurred from the viewpoint of the European integration process, with considerable impact on cross-border cooperation. The EU enlargement, almost doubling its size, has consolidated cooperation models among the new members and their neighbours; the Open Method of Coordination, subsequent to the Lisbon Strategy, represents a step forward for inter-governmentalism and the re-nationalisation of Community policies. The regulatory package for the period 2007-2013 has coined the term territorial cooperation in order to encompass the different cooperation models that emerged in the previous decade. The European Commission has launched, relatively successfully, macro-regional strategies, which provide a significant stimulus to cooperation.

This monograph aims to contribute to the debate on the future of cross-border cooperation with experiences from its past and its informal

---

15 B. Badie, *La fin des territories. Essai sur le désordre international et sur l'utilité social du respect*, Paris, Fayard, 1995.

16 G. Popescu, "The Conflicting Logias of Cross-Border Reterritorialization: Geopolitics of Euroregions in Eastern Europe", in *Political Geography*, Vol. 27, No. 4, 2008, pp. 418-438.

17 Examples related to this volume include: M. T. Bitsch (dir.), *Le fait regional et la construction européenne*, Brussels, Bruylant, 2003; O. Kramsch and B. Hooper (eds.), *Cross-Border Governance in the European Union*, London and New York, Routledge, 2004; B. Wassenberg, *Vers une eurorégion? La coopération transfrontalière franco-germano suisse dans l'espace rhénan de 1975 à 2000*, Brussels, PIE Peter Lang, 2007; M. Libera and B. Wassenberg, *L'Europe au cœur. Etudes pour Marie-Thérèse Bitsch*, Brussels, PIE Peter Lang, 2009; D. Wastl-Walter (ed.), *The Ashgate Research Companion to Border Studies*, London, Ashgate, 2011; T. M. Wilson and D. Hastings (eds.), *A Companion to Border Studies*, London, Blackwell Publishing, 2012.

18 Y. Lejeune (dir.), *Le droit des relations transfrontalières entre autorités régionales ou locales relevant d'états distincts*, Brussels, Bruylant, 2005; H. Labayle (dir.), *Vers un droit commun de la coopération transfrontalière?*, Brussels, Bruylant, 2006; C. Fernández de Casadevante Romani (dir.), *L'État et la coopération transfrontalière*, Brussels, Bruylant, 2007.

institutionalization. In order to do so, we have had the collaboration of professors from different social sciences, historians, geographers, sociologists and political scientists; this allows us to present a clear and convinced interdisciplinary approach. We also wanted to combine studies carried out by internationally renowned authors from the field of cross-border cooperation with other less known ones. This volume is structured in two parts. The first part, more theoretical, analyses general aspects of cross-border cooperation in the EU, with Euroregions as a recurring theme, including a chapter aimed at analysing the same phenomenon in North America. **L. Domínguez** and **I. Pires** analyse the historical development of cross-border cooperation, taking the symbolic years 1989/1990 as reference. In fact, it was along these two critical years when sustained and generous financial support to cooperation was initiated, especially, on the part of the European Commission with the INTERREG initiative. For some, it was about involving more actors in the multilevel arena which European governance was turning into; for others, it included the application of indicative planning formulas somewhat outdated; there are some who even point to the creation of privileged laboratories of a borderless Europe. The fact is that cooperation may have a different meaning, depending on the interests of each of the actors involved. Local and regional authorities see it as an instrument that can achieve a certain degree of centrality in the Community arena and that can also facilitate the endogenous development of its territories. The European Union understands it as a way of building European unity from the bottom, steadily, removing mutual prejudices and forging a truly European citizenship. States manage cooperation in a Lampedusan style, maintaining the control and deepening in the respective national identities, banishing the spectre of the creation of new territorial identities with relative success, except for the areas with political dominance of regionalist and nationalist forces on both sides. Finally, when cooperation structures are criticised as being mere communication tools for guidelines from Brussels, it is worth remembering that cross-border cooperation is distantly rooted in the European construction process itself and that its actors and motivations have not diverged substantially before and after 1989/1990.

Cross-border cooperation, both in Europe and North America, has a common starting point in the interests in intensifying economic and trade relations. Nevertheless, the institutionalization of these interests was substantially different. In fact, the foundational treaties of the European Communities went beyond trade relationships, creating their own institutional system with clear supranational elements and common policies which, for the particular case of cross-border cooperation, result in important financial support to the creation of cooperation structures which, in many cases, ended up forming Euroregions. Conversely,

NAFTA consciously escaped from any attempt to create similar policies or institutions; in North America, the promotion of continental trade has always been seen as a threat to national sovereignty, as a facilitator of an unwanted political integration. **E. Brunet-Jailly** highlights these differences. In North America, the support of trade relationships and the increase in security are the most important factors which explain the set up of cooperation structures across borders. Cross-border cooperation results from policy parallelism. However, the author also emphasises a second fundamental difference between Europe and America, the existence or not of a recent past of warfare, which stimulated cooperation in the case of Europe in greater deal than in North America. At the same time, each U.S. border has different characteristics. In fact, on the border with Canada, there exists a higher number of cooperation structures and the participation of the private sector is very remarkable. We could even talk about emerging proto-border regions in the West (Cascadia) and the Great Lakes area, with a special mention to PNWER. On the other hand, there are fewer cooperation structures on the border with Mexico, and nearly all of them are linked to the public sector, being this federal, regional or local. Here, the consequences of the 9/11 on security reinforcement result in a greater difficulty for the private sector to participate in these structures. Also, in contrast with the wider areas in which these structures are developed in Canada, structures with more specific aims or interests predominate here. Furthermore, while security issues are the subject of border cooperation with Canada through Integrated Border Enforcement Teams, there is no similar unit on the border with Mexico.

B. Wassenberg introduces us to the recent linkage between cross-border cooperation and international relations, a connection which has not always been evident. In fact, early academic approaches to cross-border cooperation in Europe, carried out by geographers in the 1970s from the standpoint of international geopolitics, regarded this phenomenon as part of the post World War II reconciliation process in Europe. It was not until the 1990s, with the interest in regionalism shown by political scientists, that cross-border cooperation became linked to the process of European integration; a process which promoted cooperation as a key element, much needed by the newly created single market, in order to overcome borders. However, the author argues that it is in this century when a more general connection with international relations is established and the role of cross-border cooperation is recognized as a model of micro-diplomacy.

Undoubtedly, the European integration process is experiencing a critical juncture and territorial cooperation has not only strengthened its institutional importance and become one of the three objectives of cohesion policies, established a new legal entity, namely the EGTCs, or applied their own regulations, but it can also be tackled as a reflection of this crisis, and may provide useful alternatives. The following three chapters draw attention to these aspects.

J. W. Scott claims that we can clearly observe the main elements of the EU crisis in terms of its political identity, governance, social agendas and economic perspectives through the development of Euroregions, especially those in Central and Eastern Europe. The development of Euroregions, which started at a local level as relatively modest expressions of good "neighbourliness" in Dutch-German border regions to then become laboratories of experimental governance in the 1990s, distinctly shows the process of Europeanisation followed by agendas, discourses and practices of the processes of territorial cooperation. Nevertheless, after two decades of institutional and financial support to cross-border cooperation, borders are still a major impediment to social and political interaction among citizens and institutions on both sides of the border. Moreover, cross-border cooperation results are not immediate in the regional development of the territories involved in it, nor in their economic development. Furthermore, while the Europeanisation process takes place, contradictions within the EU are contributing to a re-nationalisation of the political contexts in which local and regional cross-border cooperation is developed.

M. Perkmann focuses on the oldest Euroregion, the Dutch-German EUREGIO, which has become a symbolic reference of cross-border cooperation in the EU. Particular emphasis is placed on the analysis of both the local and supralocal context conditions under which this cross-border region emerged and the governance structures created as a result of its development, as well as an evaluation of its success and impact. The author points out three relevant aspects in the analysis of the EUREGIO experience; firstly, he sets three criteria in order to assess the success of this type of cooperation structures: organisational development, diversification of resource base, and appropriation of cross-border activities. Secondly, despite evidence of the capacity of EUREGIO and other structures to create new models of governance, their originality is shaped by the EU cohesion policies that provided opportunities for the emergence of these Euroregions. Finally, considering the evident impact of EUREGIO as a model for other EU territories, we must take into account the varied typology of Euroregions, in particular instrumental Euroregions, which are almost totally dependent on Community funds, and permanent Euroregions, which are able to develop policy entrepreneurship and have their own policies to a certain extent. EUREGIO is an example and a model of the latter group, these regions are capable of functioning as representatives of local authority interests and as implementation agencies for this type of EU cohesion policy.

Finally, **M. van der Velde** examines a key aspect which is centred on borders but clearly goes beyond their spatial framework, i.e. labour (im) mobility. This is a matter of concern for the EU and national authorities in these times of crisis due to its corollary of strengthening xenophobic drifts and the growth of extreme political ideas, which are often anti-European. The author starts off with the acknowledgement of the low levels of cross-border mobility of workers, in comparison with the United States, for example. Although he warns of the difficulties of measuring this phenomenon, the result is that it offers an exceptional character which is not relevant to the context of labour mobility in the EU. He focuses on the importance of three thresholds, the individual disposition to migrate, the location of the destination, and the route or trajectory. His analysis allows us to observe the intensification of labour mobility in the first years of the current century, not isolated from the Community policy to promote these flows as an integral part of cohesion. However, this process came to a halt in 2008, the financial and economic crisis being a decisive factor in this development. This chapter mainly analyses the cases of Denmark, Lithuania and Spain. The latter is a country which provides a good example of the changes in tendency, from being a country which hosted migrants before the crisis to being a country of emigrants at present. Nevertheless, even though cross-border mobility is still low, we can clearly detect the migrants' preference to work in neighbour countries. The current context of economic crisis does not allow for a detailed evaluation of the impact of Community policies on labour mobility.

The second part of this volume is devoted to case studies, dealing with core aspects of past and future cross-border cooperation covering most of the EU territory. Our contributors have studied the impact of structural funds aimed at stimulating cooperation in stable and old borders (such as the Iberian border) or more recent and less consolidated ones (such as the borders across the old Austro-Hungarian Empire); these funds include both the long-standing experience of INTERREG and the current cross-border Operational Programmes. The authors also research on citizen mobility, both in its specific strand of labour mobility and in its broadest sense, with all types of exchanges of people in Northern, Central and Eastern EU territories. The increasing importance of cities as cross-border cooperation agents, together with the challenges and problems entailed, constitutes an important field of study; as well as the strongly contested effectiveness of the latest structure of territorial cooperation with legal personality under Community law.

The first case study discusses the critical balance of funding, which characterises and intensifies cross-border cooperation in the EU more than in any other territory, **E. Medeiros** pays special attention to the experience of the four INTERREG, between 1990 and 2013, in "Raia

Ibérica". Community Initiatives have certainly contributed to the reduction of the barrier effect in border areas, however, this is a medium or long-term process, which requires the participation of local and regional stakeholders in building robust and sustainable cross-border networks; in this regard, the growing role of universities as promoters of cooperation initiatives is very significant as it is a sign of the maturity of cross-border cooperation in the "Raia". The Iberian example shows how the successive INTERREG programmes transformed one of the most isolated and unaltered EU borderlands into a dynamic space of cooperation intentions, through territorial networks, which include structures as varied as Working Communities, Euroregions, Eurocities, and EGTCs. However, the strategies used, which focus on a small number of objectives with the ultimate goal of reducing the barrier effect and promoting the territorial capital of the border area, are at odds with the lack of compromise of the different regional development strategies, which still prioritise national perspectives over cross-border ones.

A. Torello presents the impact of cross-border operational programmes for the period 2007-2013 between Slovenia and its neighbouring countries Austria and Italy. The author focuses mainly on infrastructure enhancement and transport policies, as key features in the cross-border cooperation process, which is still unable to overcome the traffic congestion problems at specific moments, like holiday seasons; on the environmental policy, in which a threat to sustainable development of the border can be detected as a consequence of the excessive levels of water consumption and the need for monitoring the prevention of potentially frequent natural disasters; on the asymmetrical labour markets, as a consequence of the insufficient flexibility of the Slovenian framework, and of the low competitiveness of its SMEs, and higher salaries in Austria and Italy; lastly, this chapter is devoted to the continued existence of noticeable historical, cultural and linguistic barriers, especially between Italy and Slovenia, which hinder cooperation.

The following two chapters address the issue of both stable and occasional mobility on border spaces. **G. Besier** and **K. Stokłosa** focus on the German border with Poland and Denmark. In the first case, the German-Polish border region has a long history of cross-border commuting, which was even negotiated and agreed during the period under Soviet rule, directed to large factories in the GDR, with a majority of women commuters. Since 2008, mobility has gained importance and has become truly free again, although not in the numbers originally expected by most and feared by some; this issue hints at the persistence of the stereotype of Polish slackers or at how Polish workers are received on the other side of the border. However, it may be necessary to emphasise that mobility does not only occur among Polish, but also among Germans, to the extent that the number of German workers in Poland has already outnumbered those who are employed in Austria. In the second case, in the Dutch-German border, problems are of a different nature and the number of commuters is much lower. On the one hand, the number of Germans who have crossed the Dutch border is higher, especially under the labour crisis until 2009; on the other hand, there is an ongoing debate about the "unwilling destinations of immigration" from non-Western countries that encourages the right wing populists discourse in favour of establishing restrictions to migration, which seems to be more successful on the Danish side.

In his contribution to the volume, **T. Lundén** pays attention to informal cross-border contacts in the Baltic area, a privileged cooperation space recently praised by Community institutions. This type of contact has a strong influence on identity and spatial behaviour due to the fact that most modern states try to make a nation-of-will out of its population through *inclusion/exclusion* processes, and try to make domestic contacts seem more natural than those of, for example, ethnic groups beyond the border. In most European states, boundaries are porous and feelings of loyalty may be vague or divided. Even in borderlands with an almost total ethnic division, where identity is clearly delimited, a second generation of migrants, settled on the other side of the border, may change or modify their allegiance because, as the author concludes, territorial allegiance and ethnic identity are learning processes.

Although border regions or spaces as a whole captured all the interest in the past, cities have become particularly relevant as cross-border cooperation actors in recent years. Thus, cross-border metropolises start to be the main subject of study. **F. Moullé** and **B. Reitel** draw our attention to the cross-border metropolises formed by the cities of Basel and Geneva with their urban peripheries beyond territorial boundaries. Despite their small size, Basel and Geneva are "poles of excellence" at a regional level and are well connected to global networks. The cross-border dimension in these regions brings some specificities, such as the significant increase in the level of complexity, especially in political terms – it should not be forgotten that cross-border metropolises are both spatial and political constructions. Thus, for example, the analysis of public transport networks and plans shows a high degree of integration, but it also reveals the difficulties found in building a real cross-border metropolis. The main finding is that, while using the border as a source of legitimacy, seeking being more visible to national authorities, public actors are reinforcing the ambiguous and ambivalent character of the border at the local level, without overcoming the current metropolitan asymmetry.

Symbolically, this volume concludes with a case study focused on the newly established European Groupings of Territorial Cooperation (EGTCs), a long-standing demand on the part of local and regional actors to manage cooperation, overcoming the barriers derived from the co-existence of different regulatory systems on each side of the border. **E. Soós** tackles this issue from a Hungarian perspective, one of the leading countries in Eastern Europe, and even in the EU as a whole, in the implementation of this new tool. The EU has offered its members the opportunity to establish EGTCs, bodies with legal personality, since 2006. This chapter analyses the reasons why the creation and development of EGTCs, with the participation of third countries' regional and local governments, have not been able to meet their initial expectations. The main problem lies on the concept of sovereignty. Many states, especially Serbia and Romania, still understand cooperation as part of international relations, a matter which is the sole competence of central governments. At the same time, the desire for control on the part of national governments is strengthened by the possible reawakening of regional identities. The author argues that there is a complementary relationship between local and regional democracy and cross-border cooperation; this relationship cannot function fully effectively without strong local governments with efficient local governance. The need for legal national frameworks is also pointed out here; these would allow cross-border cooperation to develop from the ratification of the Madrid Convention and the Council of Europe, and from the signing of bilateral agreements, in this case of Hungary with Serbia and Romania.

Luis Domínguez and Iva Pires
Winter 2014

Part I

EU Cross-Border Cooperation

Historical Balance and Future Perspectives[1]

Luis DOMÍNGUEZ
Universidade de Vigo

Iva PIRES
Universidade Nova de Lisboa

On 28th June 2013, the Permanent Representatives Committee approved, on the Council's behalf, the agreement reached with the Parliament on the draft regulation establishing the EU's Multiannual Financial Framework (MFF) for the period 2014-2020. The overall expenditure ceiling is, for the first time, below that of the previous period, being reduced by 3%. However, the budget item allocated to European territorial cooperation has been raised from 2.40% of the overall funds for cohesion policies, in the period 2007-2013, to 2.75%. Although this may seem a small increase, it indicates a significant change in trend. In fact, as regards heading 1b of the MFF earmarked for economic, social, and territorial cohesion, 33.87% of the total EU expenditure was agreed for 2014-2020, this heading accounted for 35.64% of the total expenditure for the period 2007-2013. Thus, it is evident that the EU has reduced its investment in economic, social, and territorial cohesion policies in general, but, at the same time, expenditure on territorial cooperation has been increased.

Undoubtedly, the strong criticisms made on the efficiency and effectiveness of cohesion policies regarding previous programmes also report on this reduction[2]. The Barca Report summarises a substantial part of

1 This paper has been done within the research project 10SEC302013PR.

2 In general terms, and leaving aside the constraints of the econometric models applied due to the difficulty in finding quantifiable data, we can mention the criticism on the lack of regional convergence, notwithstanding funds allocated to cohesion, by J. Bachtler and G. Gorzelak "Reforming EU Cohesion Policy. A Reappraisal of the Performance of the Structural Funds", in *Policy Studies*, Vol. 28, No. 4, 2007, pp. 309-326; S. Dall'erba, and J. Le Gallo, "Regional Convergence and the Impact of European Structural Funds over 1989-1999: A Spatial Econometric Analysis", in *Papers in Regional Science*, Vol. 87, No. 2, June 2008, pp. 219-244; even the negative impact on regions of Germany, Greece

them[3]. At the same time, territorial cooperation, officially known as such since 2007, has also been subjected to criticism about its development since the implementation of the Community Initiative programme INTERREG, in 1990. Within the framework of the Soviet bloc dissolution and the prospect of EU enlargement to the East, the European Parliament passed a report by Riitta Myller in 1997[4]. Cross-border cooperation was then seen as an effective tool to promote togetherness and integration among the Central and Eastern European states which were trying to find their way into the EU, but the aid schemes implemented for cooperation were considered to be muddled, the cross-border and interregional character of the actions taken was reported as unnoticeable, and the harshest criticism focused on the idea that the Commission's monitoring of the programmes and their impact was unsatisfactory, without set quantitative indicators. In that same year, 1997, the Association of European Border Regions (AEBR) published a working paper on the INTERREG initiative[5]. In the light of INTERREG I and II experiences, the arguable cross-border character of the major initiative programmes was emphasised again, concluding that these were rather national cross-border programmes attached to each other. Simultaneously, a clear distinction was made between the consolidation of cooperation structures in Northwest Europe and the obvious difficulty in building them in Southern, Central and Eastern Europe. The INTERREG programmes have undoubtedly contributed to materialising the strategic goal of removing borders, understanding these as barriers. However, it has also been strongly criticised for the administrative complexity of its processes and for the few achievements attained in its 20 years of existence, particularly for not having managed to create effective legal and management tools, for having failed

or Spain, R. Esposti and S. Bussoletti, "The Impact of Objective 1 Funds on Regional Growth. Convergence in the EU. A Panel-data Approach", in *Regional Studies*, No. 42, 2008, pp. 159-173. Among the authors who have a positive view on the effect of cohesion policies upon, for example, job creation, are R. Martin and P. Tyler, "Evaluating the Impact of the Structural Funds on Objective 1 Regions: an Exploratory Discussion", in *Regional Studies*, No. 40, 2006, pp. 201-210.

3 F. Barca, *An Agenda for a Reformed Cohesion Policy. A Place-based Approach to Meeting European Union Challenges and Expectations*, Brussels, European Commission, 2009. http://ec.europa.eu/regional_policy/archive/policy/future/pdf/report_barca_v0306.pdf [accessed on 29th July 2013]. This report builds on the concept of place-based policy, and it is based on five principles: concentrating resources; orienting grants to results; mobilising and learning; strengthening the role of the Commission; and reinforcing political checks and balances.

4 R. Myller, *Report on Cross-Border and Inter-regional Cooperation*, Committee on Regional Policy of European Parliament, 1997. http://www.europarl.europa.eu/sides/getDoc.do?pubRef=-//EP//TEXT+REPORT+A4-1997-0161+0+DOC+XML+V0//EN [accessed on 29th July 2013].

5 Association of European Border Regions, *The EU Initiative INTERREG and Future Developments*, Gronau, 1997.

to sufficiently involve private economic agents, or for not having boosted innovation and development on border regions. The predominance of the public sector raises the question of the future and sustainability of cross-border cooperation, which relies too much on subsidies, once these are ended or reduced[6].

1. 1989/90: The Frontier of Cross-Border Cooperation

This chapter about the historical lessons of cross-border cooperation in Europe will start by analysing the turning point of the years 1989-1990. It is generally acknowledged that these years represent the dawn of a new stage of cross-border cooperation, it is even considered by some as the period in which it emerged. The core element of this turning point is, undoubtedly, the financial support sustained through the INTERREG initiative. Why was it launched in these years and what was its purpose?

Rereading an old text by M. Keating has been highly stimulating, as he concluded that, in the early 1990s, the EU continued to follow the post-war approach on traditional regional policies, based on guiding national planning for the amendment of market failures in the equitable spatial distribution of the economic activity[7]. Member-states had already departed from this planning due to the lack of a multiplier effect of central investments in regional and local economies, and also to the clientelism caused by that public money. Clearly, cross-border cooperation only represents a small portion of Community regional policies, but INTERREG has also been accused of not producing that multiplier effect on borders, of exhausting the potential for job creation when the stream of financial aid runs out, and even of creating purpose-oriented structures – Euroregions – whose raison d'être is benefiting, sometimes to their advantage, from such subsidies. Furthermore, we cannot avoid mentioning the existent connection between the indicative planning established in France, the activist presidency of J. Delors as the head of the European Commission, and the NUTS, one of the last

---

6 J. Blatter, "Debordering the World of States: Towards a Multi-level System in Europe and a Multi-polity System in North-America? Insights from Border Regions", in *European Journal of International Relations*, Vol. 7, No. 2, 2001, pp. 175-209; M. Jouen (coor.), *La nouvelle Europe s'invente-t-elle sur ses marges? Coopération transfrontalière et transnationale*, Brussels, Groupement d'Études et de Recherches Notre Europe, 2001; L. O'Dowd, "The Changing Significance of European Borders", in *Regional and Federal Studies*, Vol. 12, No. 4, 2002, pp. 13-36; J. Scott, "European and North-American Contexts for Cross-Border Regionalism", in *Regional Studies*, Vol. 33, No. 7, 1999, pp. 605-617.

7 M. Keating, "Modelos de desarrollo regional", in M. Arenillas, J. Loughlin, and T. Toonen (eds.), *La Europa de las regiones. Una perspectiva intergubernamental*, Granada, Universidad de Granada, 1994, pp. 215-232.

contributions of the French language to Community parlance, nowadays monopolised by English.

The 1980s, however, represent a critical moment for the emergence of regional power in Europe. This decade also meant the revitalisation of the European Commission – which faded away once the 'empty chair' crisis had been solved – due to the large number of directives needed to implement the provisions contained in the Single Act. Theories of multilevel governance emerged in this context. Some scholars argue that the Commission has a particular interest in increasing the number of stakeholders in the Community arena, by involving local and regional authorities as direct discussion partners, in order to counterbalance the power of the Council and the states. In fact, cohesion policy challenges state-centred governance along three main lines: European institutions establish general regulations and coordinate, subnational authorities take part in decision-making processes, and the three parts are involved in a mutually dependent relationship, rather than hierarchical[8].

Therefore, we encounter a scenario characterised by the development of regional policy, established after the first enlargement and the economic crises of the 1970s, into cohesion policy; the decision about the development of the common market to the single market; the Europeanist euphoria after the collapse of the Soviet bloc and the prospects of EU enlargement to the East. And in the midst of all this are the borders. For all these reasons, Community institutions seem to have found the perfect time to launch financial support programmes for cross-border cooperation.

In addition to this, the entry into force of the Schengen agreement has, undoubtedly, a strong symbolism. Crossing gates, which were raised and lowered for each car, are opened and removed. Customs offices, which embodied Westphalian sovereignty, are abandoned, reused by other public authorities – in many cases as the headquarters of cooperation structures – or rearranged for private use. It is also true that the "borderless Europe" praised by the EU has its most prominent representation in the physical disappearance of borders[9], and that the Commission unveils new border regions as laboratories of European integration which serve as an example of overcoming national borders[10]. Border regions are, thus, on the front

---

8 L. Hooghe, "La política de cohesión de la Unión Europea y los modelos enfrentados de capitalismo", in I. Llamazares, and F. Reinares (eds.), *Aspectos políticos y sociales de la integración europea*, Valencia, Tirant lo Blanch, 1999, p. 205.

9 The absent (no-)border mentioned by O. Kramsch, and B. Hooper, "Introduction", in O. Kramsch, and B. Hooper (eds.), *Cross-Border Governance in the European Union*, London and New York, Routledge, 2004, p. 8.

10 For example, in Commission of the European Communities, *Report from the Commission to the European Parliament and the Council. Sixth progress report on economic and social cohesion* {SEC(2009) 828 final}, 2009, p. 13. The scientific literature

line of the major challenges ahead for the future of Europe, namely, the management of diversity through cohesion maintenance; they can be places for learning from the differences and for building negotiated solutions[11].

1.1. Cross-Border Cooperation for the EU

Cross-border cooperation would be, first of all, an effective tool for the EU to promote the idea of European unity, built on the principles of the Medieval Christianitas separation[12]. Secondly, it would function as a cohesive tool in territories afflicted by the disappearance of customs, its functions, and its officers. Finally, the Commission maintains that cooperation has achieved the objective of breaking down mutual prejudices and ethnical stereotypes of the other[13]. Therefore, borders are not mere economic, natural and spontaneous territories[14], they have been targeted specifically with policies, and cross-border cooperation was adopted with the aim of contributing to the European project, which seeks to build a coherent economic, social and political space[15].

With regard to the symbolic objective of overcoming barriers to mobility, it is clear the success achieved in the mobility of goods, on which the common market has been forged since the Treaties of Rome, and even citizen mobility with touristic flows in all directions, especially towards the South, flows of students through Erasmus programmes[16],

also coined this expression a long time ago, applying it, generally, to Euroregions. See M. Kessler, "Laboratory for European Integration: The Euroregions", in *Inter Nations*, Basis-Info, No. 14, 1999, p. 14; L. Roch, "Grenzüberschreitende Regionentwicklung-Basis europäischer Integration?", in *Welt Trends*, No. 22, 1999, p. 45 [44-66].

11 M. Jouen (coor.), *La Nouvelle Europe S'Invente-T-Elle sur ses Marges?, Coopération Transfrontalière et Transnationale, op. cit.*

12 J. Palma Andrés, director of DGRegio, claims that territorial cooperation "offers a unique opportunity for Regions and Member States to divert from the national logic and develop a shared space together, build ties over borders and learn from one another". See European Union, *European Territorial Cooperation. Building Bridges between People*, Luxembourg, 2011, p. 8.

13 European Commission, *Structural Policies and European Territory. Cooperation without Frontiers*, Luxembourg, 2002, p. 7. More recently, the Commission regards it as the most important achievement of territorial cooperation, as "it has helped to build trust and shared identity between people across borders". See European Union, *Territorial Cooperation. Building Bridges between People, op. cit.*, p. 10.

14 B. Jessop, "La economía política de la escala y la construcción de las regiones transfronterizas", in *Revista Eure*, Vol. 29, No. 89, 2004, pp. 25-41.

15 J. Clark, and A. Jones, "The Spatialities of Europeanisation: Territory, Government and Power in 'EUrope'", in *Transactions of the Institute of British Geographers*, Vol. 33, No. 3, 2008, pp. 300-318.

16 It is worth mentioning that the next Community arrangements for the period 2014-2020 reinforce this programme with "Erasmus for all"; among other projects, the European Parliament intends to support "Erasmus for local and regional authorities" as a way

or flows of retirees from the North towards milder climates, taking up secondary residence in the South. As far as the border is concerned, however, an important issue as labour mobility does not seem to have been achieved significantly[17]. Despite this, removing barriers to labour mobility is expected to contribute to an efficient allocation of labour and to increase the efficiency of European economy[18]. A number of reasons have been suggested to explain the low level of intra-EU labour mobility: legal and administrative barriers; the lack of familiarity with other European languages and cultures; the financial costs of moving; social security constraints, like the limited portability of pension rights; or even socio-economic and personal characteristics[19].

---

to promote exchanges of experiences and good practices of territorial cooperation. See European Parliament, *REPORT on Objective 3: a challenge for territorial cooperation – the future agenda for cross-border, transnational and interregional cooperation (2010/2155(INI))*, 11.4.2011, p. 8.

17 M. van der Velde, and H. van Houtum, "De-politicizing Labour Market Indifference and Immobility in the European Union", in O. Kramsch, and B. Hooper (eds.), *Cross-Border Governance in the European Union, op. cit.*, pp. 41-55, estimate a low 2-3% in the number of European citizens who decide to work on the other side of the border in the EU, acting with a rationality emphasised by the concept of thresholds of indifference, derived from a real representation of the national *habitus*; later on, van der Velde emphasised the meaningful importance of national restrictions with regard to labour mobility in the European East, R. Pijpers, and M. van der Velde, "Mobility Across Borders: Contextualizing Local Strategies to Circumvent Visa and Work Permit Requeriments", in *International Journal of Urban and Regional Research*, Vol. 31, No. 4, 2007, pp. 819-835; A. Niebuhr, and S. Stiller, "Integration and Labor Markets in European Border Regions", in *Zeitschrift für Arbeitsmarktforschung*, Vol. 1, 2006, pp. 57-76, have noted a greater integration of the labour market among non-border regions than in cross-border regions within the EU-15. In the case of the border between Spain and Portugal, we can observe the polarity of cross-border labour mobility through data from 2008-2010; in fact, the Autonomous Community in Spain with the highest number of Portuguese workers is the border region of Galicia, while the border district of Guarda, in Portugal, receives the lowest number of Spanish workers, see L. Domínguez, "Asimetrías y puntos de encuentro: la Raya entre España y Portugal (1991-2011)", in L. Domínguez (Coord.), *Estudio socioeconómico de la frontera entre España y Portugal*, Diputación Salamanca, Salamanca, 2013, p. 33.

18 G. Gottholmseder, and E. Theurl, "Determinants of Cross-Border Commuting: Do Cross-Border Commuters within the Household Matter?", in *Journal of Borderland Studies*, Vol. 22, No. 1, 2007, pp. 97-112.

19 S. Bouwens, "The Dynamics of Cross-Border Labour: Commuting from Dutch Part to German Part of the Euro Meuse-Rhine 1960-2000", in *Journal of Borderland Studies*, Vol. 19, 2004, pp. 135-153; G. Gottholmseder, and E. Theurl, "Determinants of Cross-border Commuting: Do Cross-Border Commuters within the Household Matter?", in *Journal of Borderland Studies*, Vol. 22, No. 1, 2007, pp. 97-112; F. Heinz, and M. Ward-Warmedinger, "Cross-Border Labour Mobility within an Enlarged EU", in *European Central Bank – Occasional Paper Series*, No. 52, 2006, p. 36; P. Huber, and K. Nowotny, "Moving across Borders: Who is Willing to Migrate or to Commute?", in

If we focus our attention on the economic impact and on the convergence of border territories thanks to the successive contributions of Community funds through INTERREG, the view of the Commission in 1996 was that INTERREG I had generated employment in infrastructures, but it was very difficult to know if those jobs would last after their completion[20]. More recently, an ESPON study has suggested that the direct impact of INTERREG, in particular cross-border cooperation programmes (strand-A), on socio-economic development is positive, though relatively small, mainly noticeable in citizens' quality of life, which improves as more simple cooperation arrangements led by NGOs or regional and local authorities are established, and it lowers when more complex structures are involved, as in the case of Euroregions, Community or national authorities or bodies, development agencies or chambers of commerce; quality of life improves when funds come from the same country or from the EU, and it diminishes when funds from national or private-public partners are involved. However, indirect impact is much more evident, as it creates a climate of trust and cooperation among actors which is particularly advantageous for the joint development of territories[21]. The empirical study on the Spanish-Portuguese border – the most benefited from INTERREG funds until 2006 and the one which had been isolated until the joint accession of these two countries to the European Communities – carried out by E. Medeiros confirms how difficult it is to assess direct impacts and the importance of indirect impacts. His conclusions actually indicate that, despite the weak correlation noted among investments from the approved INTERREG projects for this border and the evolution of the socio-economic development index, the additionality of this programme was strong, with a very positive contribution to local and regional development[22].

One of the key objectives of the EU with regard to territorial cooperation was to overcome prejudices and stereotypes of the other, with the subsequent building of trust. Progress along this path is clear; a good indicator is the significant number of town-twining partnerships,

---

Regional Studies, DOI:10.1080/00343404.2011.624509, 2011, pp. 1-20; EC, *Special Eurobarometer 337 – Geographical and Labour Market Mobility*, Brussels, European Commission, 2010.

20 Association of European Border Regions, *The EU Initiative INTERREG and Future Developments*, *op. cit.*, p. 12.

21 TERCO, *European Territorial Co-operation as a Factor of Growth, Jobs and Quality of Life*, ESPON, Luxemburg, 2012, p. 3. The authors emphasise that achieving direct impact on socio-economic development was never an objective of territorial cooperation.

22 E. Medeiros, *Efeito barreira e cooperação transfronteiriza na Raia Ibérica. Impactes territoriais do INTERREG-A*, Guarda and Lisbon, Centro de Estudos Ibéricos and Ancora Edições, 2011, p. 321.

particularly in Central and Eastern European countries severely afflicted by both World Wars[23]. However, there are also examples of the continuity of contradictory feelings on the border, which cooperation actions have not been able to eradicate. The case of the Ajuda bridge, between Alentejo and Extremadura in the Iberian peninsula, studied by J. Sidaway, is one of the most representative, and its structure, ruined since the beginning of the 18th century, is a perfect representation of the survival of national myths and territorial irredentism[24].

1.2. Cooperation and Regional Identities

The most interesting consideration we can make here, following the progressive reconciliation between border citizens, is probably that of the formation, or absence, of cross-border identities. European institutions have always expressed their opposition to territorial cooperation becoming a process of building new administrations or identities[25], but that has not prevented academic and political[26] debate. A. Paasi, prominent

---

23 Germany is the country with the highest number of town-twining partnerships, with more than 3,300, followed by France (2,500), Italy (2,000) and Poland (1,200). Although the creation of these partnerships was favoured by the integration process in the European Communities, it did not emerge as part of it, but was promoted by actors such as the Council of European Municipalities, Le Monde Bilingue or the United Towns Organisation. See A. Vion, "L'invention de la tradition des jumelages (1951-1956): Mobilisation pour un droit", in *Revue française de science politique*, Vol. 53, No. 4, 2003; J.-M. Palayrat, "De la CECA au Comité des Régions: le Conseil des communes et des régions d'Europe. Un demi siècle de lobbying en faveur de l'Europe des régions", in Mª-T. Bitsch (dir.), *Le fait régional et la construction européenne*, Brussels, Bruylant, 2003; F. Zucca, *The International Relations of Local Authorities. From Institutional Twinning to the Committee of the Regions: Fifty Years of European Integration History*, Brussels, PIE Peter Lang, 2012.

24 J. D. Sidaway, "The Choreographies of European Integration", in O. Kramsch and B. Hooper (eds.), *Cross-Border Governance in the European Union, op. cit.*, pp. 173-190.

25 In the last Swedish Presidency, in the second semester of 2009, due to the debates on macroregions driven by Sweden, it was established that the emergence of these would have to be based on, among other aspects, the principle of the 'three Nos': no new institutions, no new funding and no new legislation.

26 The issue has been very lively in Spain, for example in the creation of the Euroregion Aquitania, Basque Autonomous Community and Navarra in 1997, within the programme INTERREG 1994-1999, which lasted until 2000. In 2010, the Aquitania-Comunidad Autónoma Vasca was constituted again, with headquarters in Hendaye, but without Navarra (governed by a political party in alliance with the Spanish People's Party). In 2004, the Pyrenees-Mediterranean Euroregion was established between the Autonomous Communities of Catalonia, Aragón and the Balearic Islands, and the French regions of Midi-Pyrénées and Languedoc-Roussillon; in 2006, and up to the present time, Aragón suspended its membership due to cultural heritage disputes with Catalonia. The sharpening of the demand for sovereignty among the Basque and Catalan nationalist parties is no stranger to these political stances.

representative of a socially critical school of Political Geography, has claimed that EU discourses on regionalism have contributed to the emergence of cross-border regions which are organised by planners, politicians or business coalitions rather than created by historical processes or citizens' actions on their own[27]; many of the Euroregions created in the last two decades and the most recent Baltic and Danubian macroregions could be understood in this context. However, G. Hospers, using the regional formation model suggested by A. Paasi in the Öresund, one of the border regions always praised by the Community Authorities as an example of success, concludes that this cannot be considered a region from a symbolic, territorial or institutional perspective[28]. The studies conducted by A. Strüver[29] on the Dutch-German border region, the one with the longest tradition in cross-border cooperation, assert that prejudices and stereotypes as boundary construction practices are still active in people's minds and their everyday life. But the small survey conducted by the Euregio Rhine-Waal representatives is even more decisive in terms of new identities, the results of this survey conclude that the knowledge about the Euroregion was limited to "it has something to do with Germany" among Dutch population, and "it has something to do with the Netherlands" in Germany[30], in clear reaffirmation of their respective national identities. Other authors have focused on the importance of linguistic similarities as a facilitator for the emergence of new cross-border regional identities[31]. However, in an area which shared a language until the 14th century, as it is the case of the border between Galicia and North of Portugal, with a high degree of cross-border cooperation, citizens do not have a sense of belonging to the Euroregion[32]. In addition to this, despite strong cross-

27 A Paasi, "Region and Place: Regional Identity in Question", in *Progress in Human Geography*, Vol. 27, No. 4, 2003, pp. 475-485.

28 G. Hospers, "Place Marketing in Europe. The Branding of the Oresund Region", in *Intereconomics*, Vol. 39, No. 5, 2004, pp. 271-279.

29 A. Strüver, "Fair to Cloudy? The Netherlands and Germany alter 1945 on Display", in *HAGAR*, Vol. 3, 2002, pp. 357-360; "Significant Insignificance – Boundaries in a Borderless European Union: Deconstructing the Dutch-German Transnational Labour Market", in *Journal of Borderland Studies*, Vol. XVII, 2002, pp. 21-36; *Stories of the "Boring Border": The Dutch-German Borderscape in People's Minds*, Münster, Lit Verlag, 2005; "We are only Allowed to Re-act, not to Act: Eurocrats' Strategies and Borderlanders' Tactics in Dutch-German Cross-Border Region", in O. Kramsch, and B. Hooper (eds.), *Cross-Border Governance in the European Union*, *op. cit.*, pp. 25-40.

30 A. Strüver, "We are Only Allowed to Re-act, not to Act: Eurocrats' Strategies and Borderlanders' Tactics in Dutch-German Cross-Border Region", *op. cit.*

31 L. Topaloglou, *et al.*, "A Border Region Typology in the Enlarged European Union", in *Journal of Borderland Studies*, Vol. 20, No. 2, 2005, pp. 67-90.

32 In a survey conducted in December 2010, with the participation of 507 citizens of both NUT II, only 3% of the Spanish and 8% of the Portuguese felt identified with the Euroregion, versus other options like the municipality, the region, the state or the

border cooperation, there are territories, such as the Upper Rhine, in which the initial linguistic proximity has been reduced as a consequence of standardising the nation-states' educational policies, thus hindering the process of regional identity formation[33].

P. Virtanen[34], in an analysis of the case of Euroregions in Pomerania – former Eastern Europe – and Karelia – EU's external border – concludes that the old identity regionalism, able to create a new transnational space, has not furthered in these territories, partly because it takes time to create community awareness with cultural antecedents and common ideas, and partly because the Community message that comes with territorial cooperation is more in line with regionalisation than with regionalism, and, in any case, it has set the European transnational community as an objective, based on shared democratic values, rather than on micro-identities[35]. To summarise, cross-border cooperation has brought citizens on both sides of the border closer. It does not seem to have overcome all the existent cognitive barriers, but it has aroused interest in the neighbour and it has set the grounds for mutual trust. Except for very specific cases, authorities promoting cooperation structures have given more attention to regionalisation processes, which improve the quality of life of their voters, than to regionalist processes, which generate new identities confronted with their respective national ones. Cooperation structures created by these local and regional political actors are, to a great extent, tools to

---

European Union. See P. Ferreira, E. Vieira and L. Domínguez, "Citizens'Perceptions and Attitudes towards Cooperation in Cross-Border Regions: the Case of the Euroregion Galicia-North of Portugal", in I. Pires (ed.), *Borders and Borderlands: Today's Challenges and Tomorrow's Prospects*, Lisbon, 2012.

33 It is a fact that, according to data from the Office pour les langues et cultures en Alsace (OLCA), the number of speakers of the Alsatian dialect able to be understood by German-speaking Swiss or German citizens lowered from 91%, in 1946, to 63%, in 1995, being this almost nonexistent in the cohort of the youth. See W. Wassenberg, "Coopération franco-germano-suisse et identité regionales (1963-2007). L'identité regionale favorise-t-elle la coopération transfrontalière dans l'espace rhénan?", in M. Libera, and B. Wassemberg, *L'Europe au coeur. Etudes pour Marie-Thérèse Bitsch*, Brussels, PIE Peter Lang, 2009, pp. 141-160.

34 P. Virtanen, "Euregios in Changing Europe: Euregio Karelia and Euroregion Pomerania as Examples", in O. Kramsch and B. Hooper (eds.), *Cross-Border Governance in the European Union*, *op. cit.*, pp. 121-134.

35 L. O'Dowd maintains, from a possibilisit pragmatic perspective, that the European project is re-shaping borders, both with barriers, which hinder cooperation and integration, and with bridges, which function as facilitators. It is a complex process of forging a European identity based on cooperation and the creation of democratic borders, because we must not forget that state borders in Europe have been a sine qua non for the development of representative democracy. See L. O'Dowd, "The Changing Significance of European Borders", in *Regional and Federal Studies*, Vol. 12, No. 4, 2002, pp. 13-36.

increase both the visibility of their interests in national and Community forums, and their centrality, as they become privileged interlocutors of the European Community and the regional authorities, and even national partners for other countries. Ultimately, they are cooperation, and not integration, processes[36]. Border regions have clearly become permeable. Border boundaries have not.

2. The Origins of Cross-Border Cooperation. Multiple Factors, Converging Paths

Scholars have, on some occasions, considered Euroregions and other cooperation structures as mere means to achieve a purpose. Their only reason for being would be to replicate themselves, to attract the most funding and to keep the posts of their staff. M. Perkmann has defined them as bureaucratic entities, INTERREG "projects" rather than administrators of their budget[37]. However, Perkmann and other scholars[38] also acknowledge a value in Euroregions beyond the purpose-based idea, they see them as an effective tool for the dissemination of EU values and objectives, particularly with regard to neighbouring countries of the Community *limes*.

---

36 B. Wassenberg even stated that the proliferation of different cooperation structures in the same area, far from favouring the formation of regional identities, is a disturbing element. See B Wassenberg, "Coopération franco-germano-suisse et identité regionales (1963-2007). L'identité regionale favorise-t-elle la coopération transfrontalière dans l'espace rhénan?", *op. cit.*, pp. 155-158.

37 M. Perkmann, "Euroregions: Institutional Entrepreneurship in the European Union", in M. Perkmann, and N.L. Sum (eds.), *Globalization, Regionalization and Cross-Border Regions*, Palgrave Macmillan, Basingstoke, 2002, pp. 103-124. Soon after, the author himself suggested using the term border microregions, as he thought the term Euroregion was being used to refer to cooperation structures which were very different among themselves. See M. Perkmann, "Cross-Border Regions in Europe: Significance and Drivers of Regional Cross-Border Co-operation", in *European Urban and Regional Studies*, Vol. 10, No. 2, 2003, pp. 153-171. More recently, he has refined his typology and talks about, on the one hand, integrated micro-CBRs to refer to former border micro-regions, characterised by having structures with legal personality, smaller size than NUT II and a high level of cooperation, and, on the other hand, emerging micro-CBRs in order to refer to the new Euroregions in Eastern Europe which often cover territories smaller than NUT II, with a strong role for local powers, but with structures without legal personality and an incipient level of cooperation, very much linked to Community funds. See X. Oliveras, A. Durá, and M. Perkmann, "Las regiones transfronterizas: balance de la regionalización de la cooperación transfronteriza en Europa (1958-2007)", in *Documents d'Anàlisi Geogràfica*, Vol. 56, No. 1, 2010, pp. 21-40.

38 G. Popescu, "Geopolitics of Scale and Cross-Border Cooperation in Eastern Europe: The Case of the Romanian-Ukrainian-Moldovan Borderlands", in J. Scott (ed.), *EU Enlargement, Region Building and Shifting Borders of Inclusion and Exclusion*, Aldershot, Ashgate, 2006, pp. 35-51.

Noticeably, the agents who participate in its gestation and development, for example the Association of European Border Regions (AEBR)[39] established in the centre of the Euregio, have a different understanding of Euroregion, i.e. a structure composed of local and regional communities at a local and regional level on both sides of a national border, in some cases with a parliamentary assembly, with a permanent secretariat and a technical and administrative team, with their own budget, established as a legal entity, based on either private or public law. The purpose of cooperation is to develop a strategy, not individual measures, in order to apply joint decisions, agreed by all the actors involved at the national level, following existing procedures on both sides of the border, avoiding conflicts with regard to competences and structures. This organisation would never have a purpose-based character because it should use its own initiatives and credits, as a pre-condition to help and support other institutions.

Reality is that the term Euroregion is so appealing that, sooner or later, it is used in almost all cases, not only to define institutions, but also territories. Undoubtedly, the growing number of cooperation structures which, in the context of the enlarged EU, have opted to be called Euroregions – 58 out of the 63 recently created in Eastern Europe – has led to more confusion about the intensive character of the degree of cooperation, and also because these structures, in many cases, have not reached an implementation stage and only exist on paper[40]. N. Cornago[41] opens a new prospect when pointing out that the vagueness of the concept of Euroregion is the key to its success, despite the limited practical results obtained in the reported cases. In fact, private and public actors use it to refer to very different realities and with very diverse intentions. Thus, international organisations such as the Council of Europe, the OSCE or the EU see Euroregions as an effective tool to promote territorial cooperation, and, along with it, peace and stability. Central governments from nation-states, not without a certain degree of distrust, see them as a way of having cross-border cooperation under control without the costs that opposition would entail. Local and regional governments support Euroregions, they see them as a means to develop their economies internationally, without detriment to the social integration that goes with their performance and that offers an opportunity

39 Association of European Border Regions, *The EU Initiative INTERREG and Future Developments*, *op. cit.*, p. 32.

40 G. Tatzberger, and F. Schindegger, ESPON 1.1.3 – Final Report Part 2, Enlargement of the European Union and the wider European Perspective as regards its Polycentric Spatial Structure, Luxemburg, ESPON, 2005.

41 N. Cornago, "Un acercamiento al concepto de Eurorregión: dimensiones funcionales y economía simbólica", in F. Letamendía (ed.), *Acción colectiva Hegoalde-Iparralde*, Madrid, Editorial Fundamentos, 2006, pp. 33-52.

for symbolic self-assertion beyond state boundaries. This interpretation introduces political and identity connotations, that is the reason why it is feared by nation-states and supported by local and regional authorities with strong presence of nationalist and regionalist parties[42].

Notwithstanding the above, if we were to find a common meeting point for the academic definitions of the term Euroregion, it would be, without a doubt, cooperation. In fact, either as a technocratic instrument to develop the initiatives of European institutions, or as tools which go beyond national boundaries seeking new transfers of political loyalties, Euroregions become spaces of cooperation in which, at least, three main actors are involved: regions, with their grassroots dynamics; states, which determine the legal-administrative framework; and Europe, which provides legitimacy and financial support[43]. The territorialisation of the concept consolidated by scholars is highly relevant, in contrast with the institutional perception which prevails in the definitions of public bodies or regional associations[44], bearing in mind that when we speak about institutional perception, we do not mean political-administrative institutions, but more or less formalised structures. However, there are experts who combine both perspectives and, like E. Medeiros[45], define Euroregion as a region which comprises one or more European borders, characterised by having a strategic plan and a common vision of socioeconomic development; a territorial management structure with legal personality (like an EGTC); a high degree of shared social equipments; solid and effective cooperation in the entrepreneurial and technological domain; and a complementary urban network strongly articulated.

42 In the 1990s, at the peak of the development of multilevel governance theories and the overcoming of the nation-states framework, Euroregions were seen by some as a useful tool to demolish national states. See P. Le Gales, "Gouvernement et gouvernance des régions: faiblesses structurelles et nouvelles mobilisations", in P. Le Gales, and Ch. Lequesne, *Les paradoxes des régions en Europe*, Paris, La Découverte, 1997, pp. 237-263.

43 We have used the ideas of B. Wassenberg, *Vers une eurorégion? La coopération transfrontalière franco-germano suisse dans l'espace rhénan de 1975 à 2000*, Brussels, PIE Peter Lang, 2007, pp. 19-21.

44 However, there are also other scholars who have opted for this institutionalist perspective when defining Euroregions. It is the case of, among others, C. Olivesi, who argues that Euroregions are seen both in positive and negative terms. They are not administrative entities, nor territorial communities. They are regional associations of cross-border cooperation whose purpose is to promote close relationships on the basis of common characteristics and interests. See C. Olivesi "Du développement structurel à l'espace euro-méditerranéen: les îles et la construction européenne", in R. Balme, *Les politiques du néo-régionalisme: action collective régionale et globalisation*, Paris, Economica, 1996, pp. 207-233.

45 E. Medeiros, "(Re)defining the Euroregion Concept", in *European Planning Studies*, Vol. 19, No. 1, 2011, pp. 141-158.

But, what can a brief historical review of the period prior to 1989/90 offer to the understanding of the current phenomenon of Euroregions and cross-border cooperation as a whole? What answers can be found to questions such as the motivations of the different actors or the role of national governments?

2.1. Motivations

The first decades of territorial cooperation, its proto-history, have essentially three geographical scenarios. Germany's Western borders, the Scandinavian countries and the Alpine mountains – extending to the Pyrenees. Each one of them has its particular characteristics, but there are also coincidences. With regard to the motivations which drive actors to establish cooperation practices with their neighbours, we find a wide range of reasons. One of the first motivations was the desire for reconciliation which would provide the foundations for a lasting peace in an integrated Europe[46]. Germany's Western borders are, undoubtedly, the place where this incentive occurred more clearly; in Euregio Gronau, with the singular personality of Alfred Mozer, German-born, Dutch by choice, and a convinced Europeanist; in the Euregio Rhine-Waal, consisting of contested localities until 1960-63[47]; and in the Commaunauté d'Intérêts Moyenne Alsace-Brisgau (CIMAB), directed by Joseph Rey, mayor of Colmar, who started World War I fighting for Germany and ended up as a French soldier, member of the *Résistance* and pioneer of German-French reconciliation and cooperation process. The quest for peace starts by deepening the knowledge of the other as a useful method for overcoming the socio-psychological scars of the border[48]. But this quest was not confined to the early period and to the most contested historical borders. In fact, as far as back in 1984, the Merano Manifesto, agreed in the course of the second joint assembly of the first established Alpine Working Communities, ARGE-ALP and ALPE-ADRIA, identified peacekeeping maintenance in Europe again as an instrument to overcome disagreements between neighbours through dialogue[49].

---

46 M. Schack, *Grenzüberschreitende Zusammenarbeit an der deutsch-niederländischen Grenze*, Aabenra, Institut for Grænseregionsforskning, 1998.

47 A. Strüver, "We are only Allowed to Re-act, not to Act: Eurocrats' Strategies and Borderlanders' Tactics in Dutch-German Cross-Border Region", *op. cit.*, p. 30.

48 The objective set when Regio Basiliensis was established, in 1963, was "to remove border constraints between the various subterritories (of the region) inasmuch as they constitute an economic and socio-psychological factor". See http://www.regbas.ch/d_Information_in_English.cfm [accessed on 12th September 2013].

49 It is worthwhile to reproduce the initial paragraph of this manifesto: "The heads of government of the Länder, republics, regions, autonomous provinces and cantons from Central Europe to the Adriatic and from the Central Alpine region to the Eastern Alpine

R. Balme[50] rightly emphasised that peace and the process of European construction, particularly the customs union promoted by the Communities, opened new windows of political opportunities for territories located at the margins of decision-making processes, concentrated in the capitals of the states, to see cross-border cooperation as an opportunity to place themselves at the core of this new centrality, the Community space, which increases the value of the role played by borders in order to overcome them. Cross-border cooperation, despite the obvious slowness of the decision-making processes and the administrative hurdles to be overcome, is profitable for local and regional authorities because they can benefit symbolically from their contribution to "building Europe" and also turn the concerns of their municipal or regional problems, not only into national matters, but also European[51]. Regio Basiliensis is a good example of the quest for European centrality on the Rhine, especially in the EU-6 or EU-9, in order to overcome the peripherality of Basel in Switzerland, Freiburg in Germany, or Colmar in France[52]. In the case of Euregio Gronau, the cooperation process has undoubtedly attracted the attention of the respective national and regional authorities, on the German side, about the need for infrastructure and economy improvement in the area.

The initiators of cross-border cooperation have been mainly motivated, among other aspects, by the idea of restoring historical cultural identities, whose existence had been limited due to the presence of a border historically imposed. This motivation is often present, to a greater or lesser extent, in almost all cases, for example in The Regio Ems-Dollart[53]. Yet,

---

region declare their unwavering will to co-operate on all levels as good neighbours across borders to the best of their ability, always to keep open paths to each other, always to uphold dialogue, particularly when difficulties occur, and to do everything within their power to replace separation by mutuality so that no country shall take damage by another, to facilitate communication between the inhabitants of these countries without consideration of world view, nationality or a different social order so as to be able walk the path of co-operation and peace-keeping in Europe as good friends", see http://www.alpeadria.org/english/files/Chronology1978-2004.pdf [accessed on 28th September 2013].

50 R. Balme, "Pourquoi le gouvernement change-t-il d'échelle?", *op. cit.*, pp. 11-39.

51 Ph. Hamman, "La coopération urbaine transfrontalière ou l'Europe 'par le bas'?", in *Espaces et Sociétés*, No. 116-117, 2004, pp. 235-258.

52 As a matter of fact, we should talk about the three structures which originated the first cooperation structure with a clear cross-border character, the Regio TriRhena (1995), namely Regio Basiliensis (1963), Regio du Haut-Rhin (1976) and Regio Freiburg (1985). See B. Wassenberg, "Coopération franco-germano-suisse et identité regionales (1963-2007). L'identité regionale favorise-t-elle la cooperation transfrontalière dans l'espace rhénan?", *op. cit.*

53 L. Lindeborg, "Regional Deep Structures in the German Cultural Space", in S. Tägil (ed.), *Regions in Central Europe. The Legacy of History*, London, C. Hurst & Co., 1999, pp. 53-114.

it is convenient to clarify the empathy alleged to inhabitants on the other side of the border with respect to their neighbours, with whom they share cultural traditions, often as strong as language. In fact, B. Wassenberg uses data from a sociological study carried out in 1965 in order to show how the perceptions of the other are appropriate to defuse tensions. Swiss citizens from the Northeast are self-assured as a consequence of their economic and political situation, and they consider the population of Alsace and Baden to be inferior, feeling sympathetic to the former and mistrustful of the latter; Baden citizens, likewise, also feel very confident, they see Swiss as their peers and sympathise with Alsatians; finally, Alsatians are on the defensive and complain about the arrogance of the others[54]. However, culture, education and quality of life of border citizens have undoubtedly been prime concerns of cooperation structures since their beginnings[55].

In the 1950s and 1960s, public bodies showed a growing interest in spatial planning, which had a regional scope in all countries, as a necessary element for economic development. Borders were also part of this interest and cross-border cooperation found one of its main motivations in the coordination of the different regional development plans. The initial steps taken in the region of Saar-Lor-Lux go in this direction. In fact, the first contacts were established in 1962, during the course of the first cross-border conference on the common problems of spatial planning, which was held in Liège. It was an overall failure due to the fact that Lorrainers lacked competences within the centralised framework of the French administration at the time[56]. In the Upper-Rhine, in 1971, local authorities from Friburg, Haut-Rhin and Basel started the meetings of the Rhine Tripartite Conference with the main intention of agreeing upon issues related to spatial planning, it was followed by the Conference of Upper-Rhine Regional Planners in 1972. Moreover, this was the meeting point between the first cooperation

---

54 Data has been retrieved from P. Stolz, and E. Wiss, *Soziologische Regio Untersuchung*, cited in B. Wassenberg, *Vers une eurorégion? La coopération transfrontalière franco-germano Suisse dans l'espace rhénan de 1975 à 2007*, *op. cit.*, p. 60.

55 The eleven publications edited by the Working Community ARGE-ALP in its first decade (1972-1981) may serve as an example: four of these publications had culture and education as their main theme, three focused on healthcare, two on spatial planning and one on the local administration and the environment. See http://it.argealp.org/pubblicazioni/elenco-delle-publicazioni/elenco-delle-publicazioni [accessed on 28th September 2013].

56 R. Wittenbrock, "Les origines de la coopération dans l'espace Saar-Lor-Lux", in *Séminaires pluridisciplinaires sur la coopération transfrontalière. L'origine de la coopération transfrontalière*, Strasburg, Euro-Institut & Université Robert Schuman, 2008, http://www.euroinstitut.org/pdf/Download-Unterlagen/20080201-origine-transfron/CR-du-colloquetransfrontorigines.pdf.

structures to initiate the articulation of association networks, both in terms of the empowerment of regions and cross-border cooperation in itself. Regio Basiliensis[57] organised the International Conference on Regional Planning in Basel in 1965, which gathered a thousand experts and professionals (political authorities, university researchers and urban and spatial planning technicians)[58] and laid the foundations for the demands of the "Europe of the Regions", as well as the basis of what would later become the Association of European Border Regions (AEBR).

Environmental protection, another initial concern of cooperation structures, was very much related to spatial planning. This very sensitive issue was already present in the 1970s in the Alpine region, as it can be noted from the joint declaration which originated the Working Community ALPE ADRIA, signed on 20th November 1978, in Venice[59]. This particular aspect would remain present in the following years and it would strongly influence the articulation of the organisational network of cooperation in the area, as we can exemplify, among others, with the creation of the Espace Mont-Blanc[60]. Spatial planning and environmental protection are also linked to the motivations which led to the creation of CIMAB[61]. Ch. Ricq claims that the desire to find solutions to environmental problems explains the emergence of cooperation in the Nordic countries[62].

Concerns about the coordination of exploitation and protection of water resources shared on the borders, a factor which again links the interests of

---

57 In fact, the pioneers of cooperation on this Regio already considered that only cross-border cooperation could offer a solution to the spatial planning problems caused by the growing urbanisation of the area of Basel. See M. Lezzi, "Les suisses et la coopération transfrontalière dans la Regio", in *Séminaires pluridisciplinaires sur la coopération transfrontalière. L'origine de la coopération transfrontalière*, *op. cit.*

58 Apart from Regio Basiliensis, this conference had the participation of EUREGIO, CIMAB and the first associations of what would later become the Regio Rhein-Waal.

59 With regard to the competencies of the informal structure created, it is established, in the third article, that "the following will be dealt with: trans-Alpine traffic links, port traffic, generation and transmission of energy, agriculture, forestry, water management, tourism, environmental protection, nature conservation, landscape care, preservation of cultural and recreational landscape, regional development, settlement development, cultural relations, contacts between scientific facilities". See http://www.alpeadria.org/english/index.php?page=656102051&f=1&i=733044516&s=656102051 [accessed on 28th September 2013].

60 E. Lissandrello, "Cross-border region Espace Mont-Blanc: a territorial 'not-yet'?", in O. Kramsch, and B. Hooper (eds.), *Cross-Border Governance in the European Union*, *op. cit.*, pp. 88-103.

61 M. Anderson, "The Political Problems of Frontier Regions", in M. Anderson (ed.), *Frontier Regions in Western Europe*, London, Frank Cass and Company Limited, 1993, pp. 1-17.

62 Ch. Ricq, "La coopération dans l'espace Léman", in *Séminaires pluridisciplinaires sur la coopération transfrontalière. L'origine de la coopération transfrontalière*, *op. cit.*

territorial planning and environmental protection, have played a key role in motivating cooperation. It is no coincidence that many regions with a leading role in this process share a river, lake or coast. For instance, in the Euregio Rhine-Waal, the first step towards formal cooperation was the creation of the 'Interest Group for the Gelders River Area', in 1963[63]. In its early days, the International Bodensee Conference combined the three issues: spatial planning, environment, and water resources on Lake Constance[64]. In another territorial space analysed here, the Spanish-Portuguese border, the first contacts between regional authorities for the development of cross-border cooperation policies, which took place in 1983 between the regional government of Galicia and the authorities of the Comissão de Coordenação Regional do Norte de Portugal, centred their attention on the search for common solutions to the basin of the river Miño, which constitutes the national border[65].

Reasons for cooperation have not always been so proactive, in search of reconciliation, healing the scars of history which are borders, coordinating spatial planning, environmental protection or water resources. Occasionally, those motivations are found on reactive circumstances derived from adverse economic situations. Such is the case analysed by M. Perkmann on the origins of Euregio Gronau. Here, one of the elements of interest for actors was the search for an alternative to the problems of unemployment caused by the restructuring of the textile sector which, on the German side of the border, affected 50% of the manufacturing industry labour force and where unemployment rates reached the 80% at the beginning of the 1960s[66]. R. Wittenbrock has pointed out that the motivations which originated cross-border cooperation in the space Saar-Lor-Lux also need to take into account the hope of overcoming the economic crises derived from the progressive decline of the iron and steel industry, as well as coal mining, through cooperation[67]. Even such an emblematic landmark for cross-border cooperation in Europe

---

63 A. Strüver, "We are only Allowed to Re-act, not to Act: Eurocrats' Strategies and Borderlanders' Tactics in Dutch German Cross-Border Region", *op. cit.*, p. 30.

64 A. Hildenbrand Scheid, *Política de ordenación del territorio en Europa*, Seville, Universidad de Sevilla y Consejería de obras públicas y transportes de la Junta de Andalucía, 1996, pp. 265.

65 In fact, the issues dealt with were: the prohibition of extracting inert material, a shared dam, the preservation of fishing resources, and facilitating communications by means of bridges or ferries. See *Nota sobre a visita do Presidente da CCRN a Santiago de Compostela em 14 e 15 de Julho de 1983*, Biblioteca da Comissão de Coordenação e Desenvolvimento Regional Norte, Porto.

66 M. Perkmann, "Construction of New Territorial Scales: A Framework and Case Study of the EUREGIO Cross-Border Region", in *Regional Studies*, Vol. 41, No. 2, 2007, pp. 253-266.

67 R. Wittenbrock, "Les origines de la coopération dans l'espace Saar-Lor-Lux", *op. cit.*

as the Öresund bridge was strongly motivated by the difficulties of the shipbuilding sector towards the end of the 1980s[68].

Finally, in the Nordic countries, the dynamics of cooperation, as we will see, have other actors and rhythms, and because of that, their initial motivations and objectives also offer some particular characteristics with regard to the other spaces analysed. The Nordic Council, established in 1952, held its first session in February 1953 in Copenhagen, with the presence of the Prime Ministers of the four member countries at the time, nine recommendations were adopted in the sphere of social, economic and cultural issues, health service, communications and free flow of citizens in the Nordic countries. It is worth mentioning some of them: recognition of university studies carried out, partly, in another member-state, mobility of university professors, joint research on marine biology, abolishment of passport requirement to travel from one Nordic country to another and freedom of residence for their citizens, etc.[69]

2.2. Actors

With regard to cross-border cooperation actors, there are several elements to discern; firstly, whether actors are national or local; secondly, whether they are public or private. It is often the case that the composition is mixed. It was in the Scandinavian countries where formal cooperation, more or less institutionalised, was first articulated, and their model was clearly intergovernmental. The initiative was taken by national authorities. Governments formed inter-state cooperation bodies and, soon after, signed agreements or treaties which established a legal basis for the creation of cooperation structures between local and regional powers. The Nordic Council (between Denmark, Iceland, Norway and Sweden, with the official incorporation of Finland in 1956) was constituted in 1952 as a policy coordination body among its members[70]. It was an advisory body for parliaments and their

[68] M. Rusiñol Mussons, *La política europea de transportes: análisis de los efectos territoriales de los enlaces fijos*, Tesina del Departamento de infraestructuras del transporte y del territorio de la Universidad Politécnica de Cataluña, Barcelona, 2008, p. 59.

[69] The full text of the session overview can be seen at the Archives of the Council of Europe, Strasbourg, Nordic Council 1949/1957, 24029, it can also be consulted at http://www.cvce.eu/ [accessed on 30th September 2013].

[70] In line with the events in Western Europe at that time, the communist parties, present in the parliaments of Denmark and Sweden, also opposed the creation of cooperation bodies among countries, as they saw in them the extended arm of American diplomacy in the context of the Cold War; once the Council was established in 1953, communists would not get enough votes in any of the parliaments to send a delegate to the sessions. See F. W. Wendt, *The Northern Council: its Background, Structure and First Session*, Historical Archives of the Council of Europe, Strasbourg, Nordic Council 1949/1957, 24029, 1957, it can also be consulted at http://www.ena.lu.

respective governments which did not have executive power, its function was to advise as well as to adopt recommendations. In 1962, the Helsingfors or Helsinki Treaty[71] was signed, thus creating the legal basis for cooperation among Nordic states, they committed to cooperate at the legal, cultural and socioeconomic level, as well as in infrastructure issues[72]. The truth is that this treaty, on the one hand, reflected several of the recommendations which emerged from the first meeting of the Nordic Council, particularly those related to citizens' freedom of movement and residence, and to cooperation in the field of education and research; on the other hand, it was intended as a Northern reproduction of the parallel process of customs unification carried out by the Communities and thus, there was a mention to the liberalisation of capital movements to the highest possible extent and to the removal of customs barriers to trade and to relations between enterprises within the Northern framework (art. 18 to 25). The degree of intergovernmental cooperation was strengthened in the 1970s. While the AEBR was created in the continent as a paradiplomatic lobby or interregional interest group, the Nordic countries established the Nordic Council of Ministers (1971) as a forum for intergovernmental cooperation with decision-making processes on the basis of unanimity, as it is appropriate in this type of institutions. The Council of Ministers became the entity that would provide public funds for cooperation programmes, therefore anticipating the emergence of regional policies in the European Communities[73]. Following the same sequence we have already seen before, the creation of this institution was accompanied by a legal text, the Nordic agreement of 1977, subscribed by Finland, Norway, Sweden and Denmark, which stated that cooperation among municipalities along national borders should be easy to achieve, as if it were cooperation between municipalities of the same country, and for this to be possible, the different parties were requested to implement

---

71 This treaty has been modified several times; we would like to highlight two of them. The modification of 1983, which included Feroe Islands and Greenland (Denmark), and Aland (Finland) into the cooperation bodies of the Nordic Council; and the modification of 1993, in which, for the first time, cooperation is linked to the process of European construction.

72 J. Gabbe, V. Malchus, and H. Martinos, *Guía Práctica de la Cooperación Transfronteriza*, 3rd edition, Gronau, Association of European Border Regions, 2000, pp. B2, 16.

73 Manifestly, at present, the comparison between the funding of regional cooperation provided by the Nordic Council of Ministers and the EU is too favourable to the latter, in a proportion of one to twenty, M. Schönweitz, "The Öresund Committee. Cross-border institution-building in the Baltic Sea Region", in *NORDEUROPAforum*, Vol. 2, 2008, pp. 75-94.

the necessary changes in their national legislations[74]. It established a fundamental principle: the right of local communities to cooperate with their neighbours, except for powers reserved to national authorities[75].

In the Euroregions established on the Western borders of Germany, the initiators of cross-border cooperation were mainly local authorities. This is the case of the Euregio Gronau, constituted by three associations of municipalities, one German: *Interessensgemeinschaft Rhein-Ems* (1954), formally organised later on as *Kommunalgemeinschaft Weser-Ems* (1962); and two Dutch associations: *Belangengemeemschap Twente- Gelderland*, and *Samenwerkningsverband Oost-Gelderland*. Municipalities were, again, the main initial actors in the case of the other Euroregions in the area, as it was the case for the Euregio Rhine-Waal or the Ems-Dollart. In the Upper-Rhine, the Regio du Haut Rhin was developed with the support of the mayor of Colmar, Joseph Rey; and the Regio Freiburg was promoted by Rold Böhme, mayor of Freiburg. In the Scandinavian model, municipalities were also prominent actors under the protection of the interstate treaties mentioned above.

Regional authorities are active in the beginnings of cross-border cooperation, particularly when it comes to the creation of Working Communities with Alpine countries, and they participate in some of the German-Dutch Euregios, as it is the case of the Euregio Rhine-Waal. Also active are the entrepreneurial entities of semi-public character, like the Chambers of Commerce present in, for example, CIMAB, Regio Basiliensis, Euregio Rhine-Waal or the region Saar-Lor-Lux. Even trade union organisations are active on the French-Belgian border[76].

However, the activism of local authorities should not hide the fact that cross-border cooperation only moves forward, in these early years, when national states take control or favour its development. An institutional milestone of cooperation such as the constitution of the first cross-border Parliamentary Council in the lands of Euregio Gronau, in 1978,

---

74 In this climate, inspired by governments, the decade of 1970 saw the creation of a substantial number of cooperation structures, following the model of Scandinavian groupings: Nordkalott Council (1971), Kvarken Council (1972), Mittnorden Committee (1977), Arko Co-operation (1978), Archipelago (1978), Four Corners Co-operation (1980), Østfold/Bohuslän Committee (1980) and Nordatlantiska Samarbetet (1980).

75 The Nordic Agreement expanded cooperation to all municipalities, border or not, and among all the countries in the area, being these neighbours or not. It also established full responsibility for local powers with regard to the agreements reached, except for national governments' responsibilities, Ch. Ricq, *Manuel de la coopération transfrontalière*, Strasbourg, Conseil de l'Europe, 2006, p. 93.

76 B. Wassenberg, "Compte-rendu 1re Journée d'étude. L'origine de la coopération transfrontalière", in *Séminaires pluridisciplinaires sur la coopération transfrontalière. L'origine de la coopération transfrontalière*, *op. cit.*

was originally an idea of Prince Claus[77] – symbol of reconciliation, a former Wehrmacht officer married to the Queen of the Netherlands. The institutionalisation of cooperation between Germany and the Netherlands follows the border agreement of 1963 which, apart from the issues of delimitation, anticipated the constitution of a Permanent Boundary Waters Commission and two Sub-commissions at the regional level[78]. Similarly, in the region of Rhine-Meuse, the French and German governments created the mixed bi-national governmental Commission for cooperation in the heavy industry triangle in 1969, in the precise moment when the Saar-Lor-Lux border region was named as such. The involvement of the state was complemented by the creation of a regional forum in 1971, the regional Commission Saarland-Lorraine-Luxembourg-Trier-Western Palatinate, a model to be imitated in other border regions with French presence. The International Bodensee Conference, held in 1972, saw how the international masterplan for the region of lake Constance, approved in 1982, was the outcome of the German-Swiss Commission for spatial planning (1973) and not of the Regional Conference[79]. The impulse to institutionalisation in the Upper-Rhine came from the agreements of Bonn in 1975, which created the French-German-Swiss Intergovernmental Commission for neighbouring problems, under the authority of the Foreign Ministries of the three states. At the same time, in Scandinavian countries, the Öresund Council, established in 1964, had a slow development, until its transformation into the Öresund Committee, in 1993, due to national authorities' lack of interest: Swedish authorities were not enthusiastic about the idea of delegating central powers; on the contrary, Danish authorities were afraid of enhancing central powers given the fact that Copenhagen is located in the Öresund region[80].

Finally, the Council of Europe[81] was decisive in the beginnings of cross-border cooperation in three different ways. First, the Congress of Local Authorities in Europe (CLAE), which became the Congress of Local and Regional Authorities in Europe (CLRAE) in 1975, organised, with the assistance of the AEBR, the first two European conferences on cross-border regions, held in Strasbourg (1972) and Innsbruck (1975),

---

77 http://www.euregio.eu/sites/default/files/downloads/EUR110404%20Imagebrosch%C3%BCre%20web.pdf [accessed on 15th September 2013].

78 J.-P. Dupuy, "Legal Aspects of Transfrontier Regional Co-operation", in M. Anderson (ed.), *Frontier Regions in Western Europe*, *op. cit.*, pp. 50-63.

79 A. Hilbenbrand Scheid, *Política de ordenación del territorio*, *op. cit.*, p. 265.

80 G-J. Hospers, *Regional Economic Change in Europe: a Neo-Schumpeterian Vision*, Münster, Lit Verlag, 2004, p. 136.

81 The best study on the history of the European Council is, undoubtedly, the one recently written by B. Wassenberg, *L'Histoire du Conseil de L'Europe (1949-2009)*, Brussels, PIE Peter Lang, 2012.

respectively. Second, the European Council organised the Conference of Ministers responsible for local entities, who gathered in Paris (1975), Athens (1976), Stockholm (1978) and Madrid (1980), until they finalised the Framework Convention on cross-border cooperation between territorial entities and authorities. Finally, the Parliamentary Assembly of the European Council and the CLRAE each organised a convention; the first one, held in Galway (Ireland) between regional authorities of outermost Europe, in October 1975, concluded with a Declaration establishing the basis for the necessary polycentrism which should govern the European construction, with solidarity and territorial cohesion as the essential goal[82]; the second was about the problems of regionalisation in Europe, it was held in Bordeaux in January-February 1978 and it included, in its Final Declaration, an anticipation to what would be the theory of multilevel governance in that it defended that the necessary leading role of the regions in the European construction and in the internal affairs of their states not only meant a threat to its unity, but also strengthened it as it did not imply the impairment of local authorities' powers and competencies[83].

3. Conclusions

The years 1989-1990 represent a frontier in many respects. Between the short 20th century and our century, according to the canonical interpretation of E. Hobsbawm; between the Cold War and the present times; between the historical period of the nation-states and the post-nationalism as stated by some venturous political scientists. It also meant a speeding up in the field of European construction, reflected in Maastricht, although it was not fully achieved, except for the enlargement towards the East. With regard to cross-border cooperation, it also meant the start of ongoing economic support which has enhanced the institutionalisation of the process and its expansion across all EU borders, both internal and external.

It is important to remember that INTERREG emerged as a Community Initiative Programme within the strict competence of the European Union. Some understand this initiative as a political instrument for the Commission in order to open the dialogue with subnational actors, in

82 In the Declaration, it is said that "the request launched from Galway for a "second Europe" – the antithesis to the area of big urban republics where population, political powers and financial means are concentrated – to Europe's dominant regions, a request for solidarity and equity without which the Community or Union could not be thought of". See M. A. Clotet i Miró, *La cooperación internacional de los municipios en el marco del Consejo de Europa*, Madrid, Civitas, 1992, p. 389.

83 With regard to cross-border cooperation, in item 38 of the Declaration, there is a fervent appeal for a recovery from the effects of national border divisions, which fragment Europe.

moments of strong regional instability which coincided with the activism of the Delors Commission. However, while we do not deny the fact that local and regional actors play an increasingly prominent role in territorial cooperation, nation-states still keep control over it. Good evidence of this can be found from the very beginnings of EGTCs, which have an atypical regulation establishing the need for adapting to national legislation as if it were a Directive.

Community funding has spread cooperation structures across the European territory, most of them under the appealing designation of Euroregion, with very different characteristics; with or without legal personality, with absolute dependence on structural funds or with a certain degree of financial autonomy; with the spirit of strict technocratic bureaucracy or with the spirit of a political organisation with internal and external projection. In any case, borders aim at overcoming their condition of periphery and at achieving a certain degree of centrality within the EU arena.

Cross-border cooperation and, in general, the process of European construction have, incontestably, brought citizens on both sides of the border closer. But bringing closer does not mean closeness. That is, bringing closer does not imply willingness to share resources and public policies because, despite optimistic messages on the part of Community institutions, cognitive barriers are still present. However, the interest on the neighbour has been stimulated, and the ground for mutual trust has been laid. Except for very specific cases, when nationalist parties have a certain degree of hegemony in their region, authorities promoting these cooperation structures do so considering regionalisation processes which will improve the quality of life of their voters rather than processes of regionalism which create new identities, confronted with their respective national ones. As commented above, in the end, it is about cooperation, and not integration, processes. Border regions have become permeable, boundaries have not.

The history of cross-border cooperation shows that nation-states have never lost their control over it. The asymmetry often present on borders, from the point of view of salaries, income bracket, public policies, administrative operation, regulatory framework or competence framework works as an efficient tool for strengthening the respective national identities and for hindering processes of loyalty transfer. The perception of the EU as a technocratic structure, more or less generous with its funds, but without a soul, facilitates the instrumental vision of cooperation on the part of many of its actors, and limits their effectiveness.

Referring back to the beginning of this chapter, focusing on the period 2014-2020 and with the perspective of the previous experiences, we believe it is appropriate to formulate a lesson from the past to the future. First of all, cooperation structures have played, and will still play, an important role encouraging territorial cooperation, but they should be more proactive in the identification of matters for cooperation in each territory; they should not be mere tools for transmitting Community guidelines, but should become dynamic actors for fostering the development of their spaces, acquiring legal personality. Second, formal or institutional cross-border cooperation should go beyond the scope of public authorities and involve the civil society; this would be a good way of overcoming cognitive barriers and creating climates of trust and trading venues for enterprises. Finally, it is dangerous to link the success of cooperation to the subsidy culture which characterised the old INTERREG programmes; it is advisable to replace it, gradually, for the culture of repayment and co-responsibility, which should combine top-down strategic planning with bottom-up management, with a moderate degree of flexibility in both directions.

A Review of Cross-Border Cooperation in North America

Emmanuel BRUNET-JAILLY

University of Victoria, British Columbia, Canada

1. Introduction

This chapter is about cross-border cooperation in North America. Cooperation on the Canada/United States/Mexico borders results from expanding economic and social-cultural linkages[1]. The fundamental argument, supported by the evidence discussed below, is that trade linkages play the leading role in the development of policy parallelism and cross-border/borderland organizations spanning the Canada/USA/Mexican borders; while there are great numbers of organizations bridging the Canada U.S. borderland, the data shows, surprisingly, a lot less organizational linkages across the Mexico – U.S. border.

This is fundamentally dissimilar to the European Union experience where top-down funding and public policies enhancing regional networking have led to the creation of euro-regions linking various local and provincial (state) actors into cross-border-regional government-like organizations also called *Euregios*[2]. Because the European Union is much more than a free trade agreement it has also developed specific policies that straddle intra-European borderlands. A history of war (three between France and Germany, for instance in 1870, 1914 and 1939) have also led to a breakthrough in supra-national co-operation between Belgium, France, Germany, Luxembourg and Italy with the signature of the Treaty of Rome in 1957. This established deep institutional and policy

---

1 C. Boucher, "Toward North American or Regional Cross-Border Communities", in *Policy Research Initiative – Working Paper Series 002*, 2005, pp. 1-72; E. Brunet-Jailly, and P. Smith, "Introduction: Constructing a Cascadian Cross-Border Region", in *Canadian Political Science Review* 2/2, 2008, pp. 1-5; E. Brunet-Jailly, "Cascadia in Comparative Perspective: Canadian U-S Relations and the Emergence of Cross-Border Regions", in *Canadian Political Science Review* 2/2, 2008, pp. 104-124.

2 See E. Brunet-Jailly, *Comparing Local Cross-Border Relations Under the EU and NAFTA*, Orono, Maine, USA Canadian-American Public Policy, 2004, p. 23.

co-operation in the monitoring and production of key war industries: the *European Steel and Coal, and Atomic Energy Communities, and the European Economic Community*. When those three communities merged in 1967, they resulted in the creation of a central federative bureaucracy (the European Commission, the Council of Ministers and the European Parliament) and since then, these international government bodies have developed much cooperation and policies linking originally six but today 28 member states. Indeed, the *European Economic Community* expanded into the *European Union*, which for a number of analysts is integrating its members into a political organization that is neither a federation nor confederation of states, or, a United States of Europe. According to its president, Jose Manuel Barroso, it is the '*first non-imperial empire*'[3]. Today, the European Union (EU), and its 28 member states, form the world's second largest trading block with 16 trillion or 19% of the world trade[4]. In 1981, with the *Single European Act*, the European Union expanded its *Regional and Social policy* funds with *Community funds* including new inter-regional policies, the *Interreg funds*, and more recently the *Neighborhood policies*. *Interreg funds* have subsequently supported the development of government-like bodies linking local and regional governments across the internal borders of the Union, and their policies.

Contrary to this European experience, in North America cross-border cooperation has developed over the last 20 years thanks to a free trade agreement, the *North American Free Trade Agreement* (NAFTA), which recently became the world largest trading block, with trade valued at 19 trillion U.S. dollars, or 21% of the world GDP. Indeed, free trade has allowed Canadians and Mexicans to secure and develop their share of the U.S. market and to initiate trade with each other: in all, 93% of Canadian exports go to the U.S. and 2% to Mexico. In similar proportion, 75% of Mexican exports go the U.S., and 7% to Canada. In all, 12% of US exports go to Canada and 11% to Mexico making them respectively the first and third trading partners (China is the second largest). It is notable that trade has increased faster than economic growth; that since the signature of the *Free Trade Agreement* trade between Canada and the U.S. has expanded by about 1800%, and that trade between Mexico and the U.S. also de-multiplied by about 800% since the signature of the *North American Free Trade Agreement*[5].

---

3 http://www.youtube.com/watch?v=-I8M1T-GgRU [accessed on 26th February 2014].

4 Council on Foreign Relations: http://www.cfr.org/trade/naftas-economic-impact/p15790 [accessed on 21st January 2014].

5 E. Brunet-Jailly, "A New Border? A Canadian Perspective on the Canada-U.S. Border post 9/11", in *International Journal*, Vol. 57/4, pp. 963-978; E. Brunet-Jailly, "Forward: Beyond the Border Action Plan – A Context", in *Canada – United States Law Journal*, Vol. 37/2, pp. 273-288.

Yet, NAFTA is only a free-trade-agreement. Its role is to facilitate the expansion of tri-lateral trade between Canada, Mexico and the United States. Also, in North America because continental trade has always been perceived as a threat to sovereignty and possibly leading to political integration, NAFTA has not given birth to any large and powerful continental institutions similar to those found in Europe. On the contrary, the long-standing economic inter-dependence has grown into a fear of political integration.

Kim Nossal for instance, captured the debate between those he called the '*economic nationalists*' and the '*integrationists*' in a discourse that equated sovereignty with economic independence. As a result, the three NAFTA partners never agreed to forge a central bureaucracy to implement pan-north American policies. All matters of further economic harmonization or security or environmental standards remain primarily national in nature with cooperative mechanisms managing differences. Cooperation is the work of small separate commissions, which bring together each three state representative. Nevertheless, in this singular environment, Canada, Mexico and the United States have seen a multitude of local, regional cross-border organization develop and span their mutual borders and border regions – the data, however, points to the rise of organizational clusters across the Canada – U.S. border and fewer cross-border linkages on the Mexico – U.S. border regions.

This chapter is divided in three sections. The first two describe successively cross-border cooperation across the Canada/US and the Mexican/US border and spanning their border-regions. The third section reviews, analyses and concludes on the state and rationales for cooperation across North American border regions.

2. Economic Regions and Trans-Boundary Governance across the Canada United States Borderlands

2.1. Overview – Five Regions

The Canada – United States boundary line, known as the *International Boundary*, is the longest territorial border in the world; it runs nearly 4,000 miles or 6,500 kilometers, and including the Alaska border, which is 1,538 miles or 2,475 kilometers long, is exactly 5,525 miles or 8,891 kilometers long. Interestingly, nearly 75% of the Canadian population settled within the first 100 kilometers of the border, making Canada a borderland nation with about 25 of its 35 million people living in the borderlands. In short, despite a few very large and dense urban areas, Vancouver, Toronto, and Montreal, Canada's borderlands are sporadically populated.

Yet, there is a considerable variety of scope and type of cross-border organizations spanning the boundary along the Canada – United States borderlands. In the 'West' – Cascadia – linkages mirror trade and cultural similarities on both sides of the border[6]. The region accounts for the most profound and diversified economic and socio-cultural clusters and linkages. These are both thick and dense thanks to a larger number of organizations. The 'Prairies' – Great Plains, on the contrary, despite strong economic ties, have fewer organizational links when compared to the other four border regions. 'Ontario – Michigan' shows significant linkages across the Great Lakes – Heartland, with economic and socio-cultural organizations crossing the borderland. The Great Lakes being the heartland of the Canada / US economy – with the automobile industry at its core – it is not surprising to find many organizational linkages and even a sense of identity. Similarly 'Quebec-New England' ties are historical with very strong local community linkages. Also, strong trade relations support strong organizational linkages as well. In the 'Atlantic – Atlantica' there are historically strong economic, social, cultural networks as well despite an increasingly scarce population and economic activities.

2.2. Five Cross-Border Regions: the West, Prairies, Great lakes-Heartland, Quebec and Atlantic

2.2.1. The West – Cascadia

In the west, agriculture, tourism, forestry, and energy trade relation support strong economic relations and interdependency. Both the Washington state in the United States and the province of British Columbia in Canada have entertained competition and cooperation.

The Vancouver-Seattle cross-border (Cascadia) region was re-discovered in the 1990s as a regional response to the inability of both central governments to mediate the increasing economic interdependence of the region[7]. The term Cascadia originally refers to a bioregion, but today represents a complex ideational construct that spans social, economic, political, cultural and environmental elements, but also refers to a delimited space – that of British Columbia, Washington, and Oregon[8] at the core of which lies the Salish sea, the extensive land and sea ancestral territory of the native Salish people. The resulting

6 See Appendix A – List of Cross-Border Organization Spanning the Canada – United State Borderland.

7 S. Clarke, "Regional Transnational Discourse", in *International Journal of Economic Development; the Politics of Ideas and Economic Development in Cascadia 2/3*, 2000, pp. 360-378.

8 S. Clarke, "Regional Transnational Discourse", *op. cit.*

transnational political community legitimized a variety of multi-level governance arrangements well documented by Blatter[9], Abgral[10], and Brunet-Jailly[11]; *the Cascadian Task Force, the Main Street Cascadia, the Cascadian Economic Council*. Today, all preexisting organizations along with *the Cascadia Project, the Pacific Northwest Economic Region (PNWER), the British Columbia-Washington Council, and the Pacific Corridor Enterprise Council* (see annexe), characterize this emerging political community. All are policy or industry focused functional organizations of co-operation. It is important to note here that the role and preeminence of the private sector is noticeable. Indeed, today, PNWER is an operating public/private sector partnership designed for the facilitation of trade across the larger region[12]. At its last convention PNWER brought together well over 700 regional policy makers, elected official and business leaders to discuss the future of the region[13]. Also, the borderland region is the birthplace of the Integrated Border Enforcement Teams. The idea of borderland security networks, the Integrated Border Enforcement Teams[14] was tested at the Blaine gateway in the late 1990s[15].

---

9 J. Blatter, "Emerging Cross-Border Regions as a Step towards Sustainable Development? Experiences and Considerations from Examples in Europe and North America", in *International Journal of Economic Development 2/3*, 2000, pp. 402-439.

10 J. F. Abgrall, "A Survey of Major Cross-Border Organization between Canada and the United States", in *Policy Research Initiative Working Paper Series 009*, 2005, pp. 1-50.

11 E. Brunet-Jailly, "Cascadia in Comparative Perspectives: Canada-US Relations and the Emergence of Cross-Border Regions", *op. cit.*; E. Brunet-Jailly, *Examining the Beyond the Border Dialogue within the Contexts of International Relations and Schengen*, Border Policy Research Institute, Western Washington University, USA, 2011, http://www.wwu.edu/bpri/files/2011_Jul_BBWG_Seminar_Proceedings.pdf#page=11 [accessed in October 2011].

12 See, in particular, http//www.pnwer.org/background/history.html; J. F. Abgrall, "A Survey of Major Cross-Border Organization between Canada and the United States", *op. cit.*

13 PNWER, "Bio energy in the west, a business and policy forum – opportunities for regional Canada-US cooperation in the next wave of the bio-energy economy", Calgary, Alberta – July 20, 2010 / PNWER Annual Conference, 2011, http://www.pnwer.org/Portals/0/documents/annual%20report/2010%20Annual%20Report%20(WEB).pdf [accessed on 19th January 2014]. Alberta Government, Alberta-Blog "PNWER summit a great success for Alberta", International and Intergovernmental Relations, 2011, http://alberta.ca/blog/home.cfm/2011/7/22/PNWER-summit-a-great-success-for-Alberta [accessed in October 2013].

14 IBET, Integrated Border Enforcement Teams, http://www.rcmp-grc.gc.ca/ibet-eipf/reports-rapports/2010-threat-menace-eng.htm [accessed in October 2011].

15 E. Brunet-Jailly, "NAFTA, Economic Integration, and the Canadian-American Security Regime in the Post-September 11, 2001 Era: Multi-Level Governance and Transparent Border?", in *Journal of Borderlands Studies*, 19/1, 2004, pp. 18-37.

2.2.2. Prairies – Great Plains

A strong agricultural sector trades primarily in cattle and grain. However, the great plains share a sparse population, which results in very few cross-border organizations spanning the boundary line. The Canadian province of Saskatchewan and Montana have signed an intergovernmental agreement in 2005 to host the Mid-Western legislative conference. Manitoba travelled to Europe with the U.S. Mid-Western states. And, together the Canadian Provinces of Alberta, Saskatchewan and Manitoba are associate members of the *US Western Governors Association*. Also, the US based think tank located in El Fargo, Dakota, *the Northern Great Plains Inc.* has focused its attention on the economic development of the border region. Its recent focus has been on energy security and bio-based materials. *The Red River Basin Commission* also brings together Canadian provinces and US states to discuss ecological and economic matters.

2.2.3. Ontario Michigan – Great Lakes-Heartland

The Great Lakes-Heartland is at the very core of Canada – U.S. trade relations; nearly 50% of all trade actually crosses the border in Windsor (Ontario, Canada) – Detroit (Michigan, USA). This cross-border region is at the center of all relations thanks to the outstanding growth of the automobile industry, which in 1902 straddled the boundary to have access to the very large worldwide Commonwealth markets.

The *International Commission for the Great Lakes* (IJC) was formed in 1907. It allows both federal governments to monitor the lakes' ecosystems. The IJC manages a large and complex partnership involving all levels of government in both countries, as well as the private sector; but it is significant that over the last decades municipalities have taken over policy initiative and implementation, and lead the 'green' agenda[16].

By 1929, southwestern Ontario, Canada, and Michigan, USA, were the automobile industry world headquarters. From inception, intra-corporate trade made up a significant part of the trade flow. The 1950s AutoPact enshrined this relationship. And since 1995, Ontario and Michigan have developed a special relationship, which seems to have resulted from their economic and geographic proximity[17] as well as their similar social and cultural values[18].

---

16 I. Kusmierczyk, "Water Over the Bridge: Examining Transnational Municipal Networks of American and Canadian Local Governments in the Context of Canada-U.S. Bilateral Environmental Relations within the Great Lakes Basin", Ph.D. dissertation, Vanderbilt University, Tennessee, USA, 2010.

17 T. Courchene, *From Heartland to North American Region State*, Toronto, University of Toronto Press, 1998. See Introduction and Conclusion in particular.

18 E. Brunet-Jailly, *Cross-Border Relations in Lille (France) and Detroit (USA)*, Ph.D. dissertation, University of Western Ontario, London, Ontario, Canada, 1999, pp. 207-252;

Other trans-boundary organizations include *the Great Lakes Commission*, and *the Great Lakes Fisheries Commission* that were established in 1955; and *the Great Lakes Information Network* since 1993. *The International Association of Great Lakes and St. Lawrence Mayors* held its first meeting in 1987, *the Great Lakes Cities Initiative* was created in 2002, and *the Council of Great Lakes Governors* started meeting in 1983. Another group is *the Canadian-United States Business Association (CUSBA)*, which promotes close relations between the business communities on both sides of the border.

More recently, the business community, in particular the big three automakers GM, Ford and Chrysler, expressed concerns regarding the border infrastructures. Also, governments very successfully implemented the Integrated Border Enforcement Teams, which provide incomparable border security across the whole region[19]. And, since 2000, *the Detroit River International Cross Partnership* was set up by federal and provincial governments to decide whether a 'new' bridge was necessary. A multilevel partnership brings together all levels of government, including municipalities[20]. In 2010, the decision was stalled with Manuel Moroun, the owner of the 1929 Ambassador Bridge, actively fighting the proposal. Today, however, it is clear, that the federal and provincial governments in Canada are about to develop a new bridge across the Detroit River, which premises the development of important partnership organizations across the region.

In brief, in the Great Lakes-Heartland region, government, business and civic organizations are active and address issues on both sides of the border[21]. Prior to the terrorist attacks of 9/11, Brunet-Jailly's assessment

---

and J. Ibbitson, *Loyal No More: Ontario's Struggle of a Separate Destiny*, Toronto, HarperCollins, 2001.

19 IBET, Integrated Border Enforcement Teams, 2011, http://www.rcmp-grc.gc.ca/ibet-eipf/reports-rapports/2010-threat-menace-eng.htm [accessed in October 2011]; E. Brunet-Jailly, "NAFTA, Economic Integration, and the Canadian-American Security Regime in the Post-September 11, 2001 Era: Multi-Level Governance and Transparent Border?", *op. cit.*; E. Brunet-Jailly, "A New Border? A Canadian Perspective on the Canada-U.S. border post 9/11", *op. cit.*; E. Brunet-Jailly, "Forward: Beyond the Border Action Plan – A Context", *op. cit.*

20 J. Sutcliffe, "The Role of Local Decision Makers in the Reform of the Detroit River Border Crossing", in *MPSA Annual National Conference*, Palmer House Hotel Chicago IL, 2008.

21 J. Nelles, "Cooperation in Crisis? An Analysis of Cross-Border Intermunicipal Relations in the Detroit-Windsor Region", in *Annual Meeting of the American Association of Geographers*, Washington DC, 2010; J. Blatter, "Emerging Cross-Border Regions as a Step Towards Sustainable Development? Experiences and Considerations from Examples in Europe and North America", *op. cit.*; J. F. Abgrall, "A Survey of Major Cross-Border Organization Between Canada and the United States", *op. cit.*

that the "*competitive city paradigm... best explains the Ontario-Michigan border region*"[22], was correct, but the post 9/11 increase in relations has led to the need to reexamine these relationships in terms of cooperation.

2.2.4. Quebec – New England

Quebec, New Hampshire, Maine and Vermont have old and strong economic linkages. Quebec also supplies electricity to New York State. The creation in the 1970s of *the Conference of Governors and Eastern Canadian Premiers* strengthened those relations in particular in the face of possible energy crises. The conference also brings its members together on environmental and economic matters with environmental cooperation for the Lake Champlain area.

2.2.5. Atlantic – Atlantica

Atlantic Canada and the northeastern states of the United States have fewer linkages than for instance in Great Lakes or the West. One important reason is that those vast border regions are less densely populated and remain primarily fishing and agrarian communities. The prominence of the conference of New England Governors and Eastern Canadian Premiers has served to exchange information but has also been criticized for not being able to address effectively the needs of their respective populations.

2.3. Summary of Findings

All in all, these short case studies point to numerous linkages and '*thickness and intensity*' across two regions that are central to the Canada-US relationship: the West and the Great Lakes-Heartland. The density is remarkable, as well as the varied scales and the functional focus of the networks that are spanning the border regions. These are networks of public and private sector managers that discuss, negotiate and implement agreements with views to secure trade, improve infrastructures or environmental standards such as water quality, and bio-energy – all assuming it would facilitate trade. Indeed, the lower level '*thickness and intensity*' observed by Brunet-Jailly[23] in the Great Lakes – Heartland has been replaced by a cluster of networks, all straddling the cross-border region, whose core is the Windsor-Detroit area. In the West – Cascadia region, the effectiveness of the local networks is best illustrated by the

---

[22] E. Brunet-Jailly, "Globalization, Integration, and Cross-Border Relations in the Metropolitan Area of Detroit (USA) and Windsor (Canada)", in *International Journal of Economic Development*, No. 2, Vol. 3, 2000, pp. 379-401.

[23] E. Brunet-Jailly, S. Clarke, and D. Van Nijnatten, "Leader Survey on Canada – US Cross-Border Regions, An Analysis", in *Ottawa, Canada: Policy Research Initiative Working Paper Series 12*, 2006, p. 7.

outstanding success of PNWER, a cross-border partnership organization that engages in varied multi-scalar policy discussions[24].

3. Economic Regions and Trans-Boundary Governance across the Mexico – United States Borderlands

3.1. Overview – Four Historical Regions – Very Few Cross-Border Organizations

The United States – Mexico border is about 2,100 miles long or 3,145 kilometers long. It is made up of four major cross-border regions: Greater California (Baja California, Mexico and California, United States), Primeria Alta (Sonora, Mexico and Arizona, United States), El Camino Real Region (Chihuahua, Mexico and New Mexico and western Texas, United States), and Nuevo Santander (Coahuila, Nuevo Leon, and Tamaulipas in Mexico, and Texas in the United States)[25].

Historically, those regions were Hispanic and Mexican before the *Gadsden Purchase* of 1853 when the United States bought California, Arizona, New Mexico and parts of Texas. Today, thanks to growing immigration from Mexico and across the southern states of the U.S., those regions have become 'Hispanic' again, with about 35% of their residents of Hispanic origin, with a majority of Mexicans. In sum, the border region is made up of about 85 million residents, 50% of which are of Hispanic origin, and it is only sparsely populated in specific areas, while others are as densely populated as any comparable urban centers[26].

There are a few government-like organizations and civic groups that operate across the U.S. – Mexico border. Since the signature of the *North American Free Trade Agreement*, a number of federal, state and even local government organizations and programs attempt to stimulate cross-border co-operation and to solve common problems.

These cross-border organizations can be divided into inter-governmental organizations – associations and agreements among states

---

24 PNWER, Bio Energy in the West, a Business and Policy Forum – Opportunities for Regional Canada-US Cooperation in the Next Wave of the Bio-energy Economy Calgary, Alberta – July 20, 2010 / PNWER Annual Conference http://www.pnwer.org/Portals/0/documents/annual%20report/2010%20Annual%20Report%20(WEB).pdf [accessed on 19th January 2014].

25 See Appendix B – List of Cross-Border Organization Spanning the Mexico – United State Borderland.

26 The following section benefited greatly from T. Payan's early writings for E. Brunet-Jailly, T. Payan, and G. Sawchuk, *The Emergence of Cross-Border Regions along the Mexican – US Border and in Europe – Lessons for Canada*, Policy Research Initiative Working Paper Series 35, pp. 1-69.

and even national governments – and non-governmental organizations, which are often managed by civil society groups.

Some cross-border organizations are bi-national, while others have a distinctively regional character: in all there are ten such organizations in Greater California, seven in Camino Real, five in Primeria Alto, and one across the Nuevo Santander border region.

Interestingly, most of these are single purpose organization with a very specific policy focus: agriculture, education, energy, environment, cultural-religious, security, health care, human rights, immigration, technology, infrastructure, sustainable development, tourism, trade, transportation, water, and workforce issues: Appendix B of this chapter details whether those organizations are national, regional or city/local, as well as their area of policy specialization.

3.2. Presentation of Select Organizations

Some organization result from State to State partnerships. On the Mexico – United States borderlands these are the NAFTA, *the Border Governors Conference, the North American Development Bank*, a non-governmental organization *the Border Trade Alliance*, and a United Nations' sister organization, *the Pan-American Health Organization.*

3.2.1. National Level Organizations Straddling the Border Region

The overarching organization, which has had the deepest effect on the border, is the *North American Free Trade Agreement* (NAFTA). Trade between Mexico and the United States has grown to nearly $500 billion since the agreement was implemented in 1994. NAFTA is ever present in the minds of nearly every organization along the border. Many are formed to take advantage of the trade agreement; others emerged as a response to the stress placed on border infrastructures, or human rights practices, environmental concerns, and many other issues. In the end, as suggested in this section, the NAFTA was determinant in the creation and development of most cross-border linkages and concurrent organizations.

Among bi-national government-like cross-border organizations, the *Border Governors Conference*, may be prominent in providing leadership to address border related issues. The governors of the Border States met for the first time in 1980 to establish a formal process of communication among the four U.S. border states (California, Arizona, New Mexico and Texas) and their six Mexican peers (Baja California, Sonora, Chihuahua, Coahuila, Nuevo León and Tamaulipas). Every year, the governors meet and discuss many issues, including agriculture, border crossings, education, economic development, energy, environment, health, tourism, and security.

A second central organization is *the North American Development Bank* (NADBANK). It promotes environmental co-operation across the border. The NADBANK is a U.S. – Mexico organization that works exclusively with local governments to fund environmental projects. Much of the work requires cooperation between cross-border partners. This is the work of the *Border Environment Cooperation Commission* (BECC), which is also in charge of the endorsement and accreditation of all projects.

The Border Trade Alliance was set up in 1986 to help its members deal with trade, travel and security issues in North America. It is located in the southern USA but acts across all three NAFTA member states. BTA represents about 4.5 million members: public and private, including chambers of commerce, industry representatives, academia, as well as local and state provincial departments.

The Pan-American Health Organization (PAHO) was set up in 1902 to provide health related technical resources to its states partners across the Americas. PAHO is an agency related to the United Nations agencies. It is very active in borderland regions in particular with regards to border health issues such as vaccination, obesity, diabetes and other related health concerns.

3.2.2. Local and State Level Organizations Straddling the Border Region

Other organizations have a strong local, regional or State partnership. These often straddle the borderland in a manner similar to that described regarding the Canada – United States borderlands. Examples include: *the San Diego Dialogue*, *the Arizona-Mexico Commission*, or *the Internal Business Accelerator* among others.

The San Diego Dialogue is especially significant because it organizes an ongoing discussion between the civic leaders of the San Diego-Tijuana cross-border region. Typical issues tackled by *the San Diego Dialogue* are region wide.

Similarly for the Arizona-Mexico region *the Sonoran Desert Corridor* provides coordination and a policy-making forum.

In the New Mexico-Chihuahua border region, *the New Mexico International Business Accelerator* promotes and supports economic development and trade.

3.3. Summary of Findings

All in all, there were very few cross-border or transnational organizations that were active in the southern border prior to the signature of NAFTA. Since the signature, however, a number of cross-border organizations emerged to resolve border linkage problems in specific areas of trade, economic development, health, migration, and human rights issues.

These organizations are nearly always specialized uni-functional organizations that represent the very specific interest of an industry, a set of government or civic group. There are more such cross-border organization in the San Diego/Tijuana region than across the other three border regions – Camino Real, Primeria Alto, and Nuevo Santander.

Most have activities that overlap the boundary but they serve very specific constituencies. Finally, when public or non-profit most of them also serve specific set of policy interests that identified by government, local, state or federal level government. Civil society remains under-represented by cross-border organizations. Unless they serve specific economic or trade issues, those are generally short lived and underfunded. In the post 9/11 era the increasing centralization of security into federal departments has further weakened the capacity of civic organizations to influence policy. Similarly, non-governmental organizations suffer from a lack of resources and leadership spanning the border. Typical examples abound with regards to environmental issues. What is most visible in the media – lacks both human and financial resources to be effective in bending policy decisions on the medium and long terms. However, protest groups have short-lived impacts such as blocking border gates at ports of entry, which are very effective in disturbing trade but are very unpopular with businesses. In the end, the most specialized and the better-networked organizations or policy issues are also the ones that are the most successful (for instance focused on trade or aspects of trade, health, environmental degradation or human right issues).

4. Conclusion

Canada, Mexico and the United States form a highly integrated economic space because of their history, because of the signature of the Free Trade Agreement and the North American Free Trade Agreement, and because trade integration has progressed at a faster rate than economic growth.

In North America, trade and security are the two primary drivers for cooperation. Cross-border cooperation therefore takes a form of specific and functionally focused approaches found in their common history of peaceful relations: functions do not spill over into overlapping policy arenas. Cross-border organizations and cooperation follow principles where both security and regulatory alignment are achieved only when necessary to specific groups or interests and are useful to the expansion or deepening of trade. For instance, specifically regulatory matters focus on the specific needs of specific industries only.

There is much evidence that cross-border regions are emerging across Canada's west and Great Lakes-Heartland regions, and there are steady relationships across the other regions of the Canada – U.S. borderland. On the Mexico – United States borderlands, despite a history and presence of increasingly dominant Hispanic culture, and very strong trading relations, there are much fewer instances of cross-border organizations linking both sides or leading to the development of cluster of networks spanning each of the four Mexican – United States border regions. The exception is the *Greater California* border region where a small number of organizations work on both sides of the international boundary line. However, their work is notably more difficult since 9/11.

To sum up, in North America, cross-border cooperation is narrowly understood as a functionally driven necessity: as trade flows and interactions spanning the borders expand, cross-border organizations are designed to facilitate those relationships. They are the managerial instruments of cooperation and policy parallelism. In some areas, such as the Cascadia or the Great Lakes – Heartlands and to a certain extent across the Greater California regions they are interestingly dense, yet their scales vary greatly and according to functional needs. They are networks of that organized policy cooperation, collaboration and result in policy parallelism. These, however, do not have the political clout and integration drive known in the European Union's *Euregios*. In North America, cross-border relations result from policy parallelism, that is, clusters of functional linkages of policy of varied scales, which purpose is to resolve managerial issues arising from human rights, immigration, security, trade and economic integration.

5. Appendix A

List of cross-border organisations spanning the Canada – United States Borderlands.

The West

Single-purpose intergovernmental:

- B.C.-Washington Environmental Cooperation Council,
- Other Task Forces (Abbotsford-Sumas Aquifer, Nooksack River Flooding, Lower Fraser Valley/Pacific Northwest Airshed, Georgia Strait, Puget Sound),
- Pacific States-B.C. Oil Spill Task Force,

- Okanagon River Basin, Western Interstate Energy Board (energy arm of Western Governors' Association),
- Western Association of State Highway and Transportation Officials,
- Western Legislative Forestry Task Force,
- Coutts/Sweet Grass,
- Western Interstate Energy Board (energy arm of Western Governors' Association),
- Western Association of State Highway and Transportation Officials,
- Western Legislative Forestry Task Force,
- Alberta-Montana Government Exchanges.

Multi-purpose intergovernmental:

- Western Governors' Association,
- Pacific Northwest Economic Development Council,
- Western Governors' Association,
- Pacific Northwest Economic Development Council,
- Specific State Relations (Montana-Alberta Bilateral MOU and Advisory Council, and MOUs with Alaska, Arizona, California, Idaho, Oregon, Texas and Washington).

Civil organizations (can include significant governmental):

- Pacific Northwest Economic Region,
- Pacific Corridor Enterprise Council,
- Cascadia Institute,
- Discovery Institute,
- International Mobility Trade Corridor,
- Northwest Environment,
- People for Puget Sound,
- Kettle Range Conservation Group.

Cities:

- Cascadia Mayors Council,
- Vancouver-Seattle (security),
- Lethbridge and Grand Falls (Montana)

Prairies – Plains

Single-purpose intergovernmental:

- Red River Basin Commission,
- International Souris River Board.

Multi-purpose intergovernmental:

- Accord with Montana,
- Western Governors' Association,
- Partnership Agreement with Minnesota (Bioscience Technology Corridor).

Civil organizations (can include significant governmental):

- Northern Great Plains Inc.,
- Great Plains Institute.

Ontario – Michigan – Great Lakes – Heartlands

Single-purpose intergovernmental:

- Great Lakes Commission,
- Great Lakes Environmental Directory and Great Lakes Aquatic Habitat Network and Fund.

Multi-purpose intergovernmental:

- Council of Great Lakes Governors,
- MOUs with New York and Michigan.

Civil organizations (can include significant governmental):

- Council of Great Lakes Industries.

Cities:

- International Association of Great Lakes and St. Lawrence Mayors,
- Great Lakes Cities Initiative,
- Toronto and Indianapolis,
- Toronto and Chicago,
- Niagara and Buffalo.

Quebec – New England

Single-purpose intergovernmental:

- Great Lakes Commission,
- Environmental cooperation with Vermont and New York regarding Lake Champlain,
- Champlain/Lacolle Port of Excellence.

Multi-purpose intergovernmental:

- Conference of New England Governors and Eastern Canadian Premiers,
- Council of Great Lakes Governors,
- MOU with New York.

Cities:

- Quebec City and Albany (New York).

Atlantic – Atlantica

Single-purpose intergovernmental:

- Gulf of Maine Council on the Marine Environment,
- St. Croix International Waterway Commission,
- Calais/St. Stephen.

Multi-purpose intergovernmental:

- Conference of New England Governors and Eastern Canadian Premiers,
- MOU with Maine,
- Atlantica: the International Northeast Economic Region.

Cities:

- Edmunston and Madawaska (Maine).

6. Appendix B

List of cross-border organisations on the Mexico – United States Borderlands.

6.1. Inter-Governmental Organizations

6.1.1. Single-purpose

National Level

- Environment Protection Agency *(Environment)*,
- Border Environment Cooperation Commission *(Environment)*,
- North American Development Bank *(Environment)*,
- Bureau of primary health care *(Health care)*,
- United States Mexico Border Health Commission *(Health care)*,
- Pan-America Health Organization *(Health care)*,
- Department of Housing and Urban Development (*Infrastructure*),
- North American Free Trade Agreement (*Trade*),
- Transportation Research Board (*Transportation*),
- International Boundary Water Commission (*Water*).

Local and State level organizations straddling at least one cross-border region:

- Border Governors Conference (California, Primeria Alta, Camino Real, and Nuevo Santander – *Environment*).

Greater California:

- San Diego Dialogue (Education, Health Care, Trade),
- San Diego Association of Governments (California – Environment),
- Frontera Audubon Society (California – Environment),
- American Society for Quality (Greater California – Transportation).

Primeria Alta:

- Arizona-Mexico Commission (Primeria Alta – *Trade*),
- Sonoran Borderlands (Primeria Alta – *Tourism*).

Camino Real:

- Energy Resources (Camino Real),
- International Business Accelerator (Camino Real – *Trade*),
- Joint Advisory Committee for the Improvement of Air Quality (Camino Real – *Environment*),
- Border Network for Human Rights (Camino Real – *Health Care*).

6.1.2. Multi-purpose

National level

- United States Mexico Cooperative Agreement (*Agriculture*).

6.2. Civil organizations

6.2.1. Single-purpose

National level

- Sierra Club (*Environment*),
- Coalition for Justice in the Maquiladora (*Human Rights*),
- U.S.-Mexico Chamber of Commerce (*Trade*),
- Border Trade Alliance (*Trade*),
- Religious Task Force on Central America & Mexico (*Culture*).

Local and State level straddling at least one cross-border region

Greater California:

- Bilateral Safety Corridor Coalition (Greater California – *Human Rights*),
- American Friends Service Committee (Greater California – *High Technology*),
- The Border Peace Presence (Greater California – *Human Rights*).

Camino Real:

- Border Philanthropy Partnership (Camino Real – *Transportation*),
- Paso Del Norte Health Foundation (Camino Real – *Immigration*),
- Leadership McAllen.

The Place of Cross-Border Cooperation in International Relations

Birte WASSENBERG

Institute of Political Studies, University of Strasbourg

Cross-border cooperation has only recently become a field of research in International Relations[1]. It is only since the late 1990s that historians have studied relations between border regions, mainly in the context of comparative "area studies"[2]. Cross-border cooperation has somehow been situated at the intersection between the history of European integration and the more general research field of border studies.

For a long time, cross-border cooperation between local and regional actors was not recognised as a tool for micro-diplomacy serving national, European or international interests, as most disciplines in the human sciences were attached to the principle that foreign policy is reserved for the nation state. It was the geographers who started to examine the geopolitical functions of borders in an international context in the 1970s[3].

Research on cross-border cooperation as a phenomenon in International Relations has suffered many impediments. First, most of the studies led by Europeans focused either on specific European border regions, i.e. on the local and regional history, geography and regional governance of a determined space within Europe, or they were designed to examine the link between cross-border regions and the European integration process

---

1 B. Wassenberg, "Historiographie de la coopération transfrontalière", in *L'approche pluridisciplinaire de la coopération transfrontalière. Les jalons pour un travail de recherche interdisciplinaire*, Cahiers FARE No. 5, 2013, pp. 9-15.

2 M. Libera, "Les historiens et la coopération transfrontalière: une rencontre tardive mais fructueuse", in *L'approche pluridisciplinaire de la coopération transfrontalière. Les jalons pour un travail de recherche interdisciplinaire*, *op. cit.*, pp. 68-74.

3 For example: C. Raffestin, and P. Guichonnet, *Géographie des frontières*, Paris, PUF, 1974; cf. H. van Houtum, "An Overview of European Geographical Research on Borders and Border Regions", in *Journal of Borderlands Studies*, Vol. 15, No. 1, 2000, pp. 57-83.

post 1945[4]. Secondly, only very few non-European researchers have undertaken work on cross-border cooperation and consequently, only few comparative studies between European and non-European border regions in America, Africa or Asia were introduced in the late 1990s[5]. It is only since 2000 that cross-border cooperation has become an interdisciplinary research subject in International Relations, thanks to the interest of the political scientists and historians who have started to specialise in that field[6].

1. Cross-Border Cooperation – a European Research Field Reserved for Local History and European Integration Studies

Research on cross-border cooperation was first launched in Europe by specialists in geography and spatial planning[7]. In the 1970s, within the Council of Europe, a number of conferences, the so-called "European confrontations", were organised in order to examine the specific problems European border regions were confronted with, especially with regard to their geographically disadvantaged position at the periphery of their respective nation states. The results were published by the Council of Europe and constituted a starting point for various researchers to continue working on the phenomenon of border regions[8]. Viktor von Malchus, a member of the Association of European Border Regions (AEBR), thus published a book, *Partnerschaft an europäischen Grenzen*, which was the first work specifically dedicated to European border regions and their spatial planning problems[9]. At the same time, a multidisciplinary university group led by the sociologist Raimondo Strossoldo started research work in Italy (Gorizia) with the objective of measuring the scope of existing cross-border cooperation

---

4 B. Wassenberg, "La coopération transfrontalière comme objet de recherche sur l'histoire de l'intégration européenne", in B. Joachim, and B. Wassenberg (eds.), *Grenzüberschreitende Zusammenarbeit leben und erforschen (Bd. 5): Integration und (trans-)regionale Identitäten*, Stuttgart, Steiner Verlag, 2013, pp. 73-101.

5 Cf. for example: J. Scott *et al.*, *Borders and Border Regions in Europe and North America*, San Diego, State University Press, 1997.

6 J. Beck, "Les enjeux d'une approche interdisciplinaire de la coopération transfrontalière en Europe", in *L'approche pluridisciplinaire de la coopération transfrontalière*, *op. cit.*, pp. 16-24.

7 Cf. for example: A. Mozer, "Entwicklungsregionen in der EWG", in *Schriften der Bundeszentrale für politische Bildung*, Bonn, 1973, pp. 14.

8 Conseil de l'Europe, *Frontier Regions and Regional Planning*, Strasbourg, 1973.

9 V. F. von Malchus, *Partnerschaft an europäischen Grenzen*, Bonn, Europa-Union Verlag, 1975.

in Europe[10]. The first approaches to cross-border regions were therefore limited to Western Europe and were more specifically focused on the analysis of the negative effects of national borders. European borders were identified as "barriers" or, worse, as "scars of history" where negative memories of wars, loss and sufferance accumulated[11]. This was generally seen as a European phenomenon, generated especially by the experience of two devastating world wars, but also by the numerous border conflicts that had opposed ruling powers in the long "historical" past of the European continent[12]. There was thus no apparent link between this research field and the more general studies of International Relations from the 20th century on.

Furthermore, in the 1980s, the first case studies on cross-border regions in Europe were published, narrowing down even more the geographic scope of research, mostly to those border regions in Western Europe which had started cooperation in the late 1950s or at the beginning of the 1960s. Empirical research was led for example on Franco-German cross-border cooperation (Saar-Lor-Lux)[13] or on the Swiss border regions (between Geneva and Basel and around Lake Constance)[14]. These case studies enabled a detailed analysis of the effects of cooperation on the border and on the local and regional actors concerned, taking into account the historical, geographical and political context of the specific cross-border space. However, being limited to the local specificities of the given border regions, these studies revealed cross-border cooperation as a research field designed for specialists in local and regional European politics, geography, sociology or history[15]. It became almost impossible to insert the very singular analyses of specific

10 R. Strassoldo, *Boundaries and Regions*, Trieste, International Sociological Institute de Gorizia, 1973; R. Strassoldo, and G. Delli Zotti (dir.), *Cooperation and Conflict in Border Areas*, Trieste, International Sociological Institute de Gorizia, 1982.

11 J. Elvert, "Über Grenzen- ein Vorwort", in B. Wassenberg, and J. Beck (eds.), *Vivre et penser la coopération transfrontalière (Volume 4): les régions frontalières sensibles*, Stuttgart, Steiner Verlag, 2011, pp. 11-25.

12 B. Wassenberg, "Qu'est-ce une region transfrontalière sensible?", in B. Wassenberg, and J. Beck (eds.), *Vivre et penser la coopération transfrontalière (Volume 4): les régions frontalières sensibles, op. cit.*, pp. 67-89.

13 W. Brücher, and P. R. Franke (eds.), *Probleme von Grenzregionen: Das Beispiel Saar-Lor-Lux-Raum*, Saarbrücken, Beiträge zum Forschungsschwerpunkt der Philosophischen Fakultät der Universität des Saarlandes, 1987.

14 C. Ricq, *Les régions frontalières de Genève à Bâle, analyse institutionnelle*, Bern, Programmleitung Nationales Forschungsprogramm No. 9, 1981; D. Bullinger, *Grenzüberschreitende Zusammenarbeit in der Regionalpolitik. Theoretische Ansätze und ihre Bedeutung für das Bodenseegebiet*, Konstanz, Verwaltungswissenschaftliche Diplomarbeit, 1977.

15 B. Wassenberg, "Historiographie de la coopération transfrontalière", *op. cit.*, p. 10.

European border regions into a more general approach on European studies, let alone on International Relations. Besides, when researchers in law started to work on the legal problems of cross-border cooperation in the late 1980s, these studies were also limited to European aspects, or more specifically to the incompatibilities between the respective national legal systems of the cooperating regions. Starting from the analysis of the European Madrid framework convention on cross-border cooperation of the Council of Europe[16], several German specialists in law (Ulrich Beyerlin, Mathias Oehm, Birgit Schlögel for example) published books on *the legal problems of cross-border cooperation*[17]. For their part, the sociologists, for example Uwe Fichtner, started to work on issues related to regional identity in a given cross-border space[18].

This singular European approach to cross-border cooperation continued in the 1990s and was generally followed by most disciplines in the human sciences: economists like Remigio Ratti or Shalom Reichman developed theories on the beneficial effects of cross-border cooperation in the specific European context of the suppression of economic borders by the Single European Market[19]. Joining the economists, the sociologists increasingly examined the socio-economic aspects of cross-border cooperation, like Markus Perkmann in his book *Cross-border Cooperation, Euroregions and the Governance of Cross-Border Economies*[20]. Conversely, the research work in law became more and more specialised within its own discipline. Thus, new legal cooperation tools such as the Karlsruhe intergovernmental agreement or questions such as conferring legal rights to cross-border institutions were discussed in detail[21]. Even the first handbook on cross-border

---

16 Council of Europe, European Series of Treaties No. 106, Strasbourg, 21st May 1980.

17 U. Beyerlin, *Rechtsprobleme der lokalen grenzüberschreitenden Zusammenarbeit*, Berlin, Springer, 1988; M. Oehm, *Rechtsprobleme Staatsgrenzen überschreitender interkommunaler Zusammenarbeit*, Dissertation, Universität Münster, 1982; B. Schlögel, *Die grenzüberschreitende interkommunale Zusammenarbeit*, Berlin, Springer, 1982.

18 U. Fichtner, "Grenzüberschreitende Verflechtungen und regionales Bewusstsein in der Regio", in *Schriften der Regio 10*, Basel, 1988.

19 R. Ratti, and S. Reichman, *Theory and Practice of Transborder Cooperation*, Basel, Frankfurt, Helbing & Lichtenhahn, 1993; R. Ratti, and M. Baggi, *Strategies to Overcome Barriers: Theoretical Elements and Empirical Evidence*, Bellinzona, Instituto di ricerche economiche, 1990; R. Ratti, *Théorie de développement des régions-frontières*, Fribourg (Suisse), Centre de recherche en économie de l'espace, 1991.

20 M. Perkmann, *Cross-Border Cooperation, Euroregions and the Governance of Cross-border Economies*, Working Paper, Lancaster University, 1997.

21 N. Levrat, *Le droit applicable aux accords de coopération transfrontalière entre collectivités publiques infra-étatiques*, Genève, Paris, PUF, 1994; J. Schoch, *Rechtliche Aspekte grenzüberschreitender Zusammenarbeit*, Bern, Peter Lang, 1997; A. Beck, *Die Übertragung von Hoheitsrechten auf kommunale grenznachbarschaftliche*

cooperation in Europe, published by the Council of Europe in 1996, contained mostly information on the existing legal tools, intergovernmental agreements or examples of cross-border regions and institutions in Europe[22]. No link with International Relations was established, although the Council of Europe had sought from its start (as early as the 1950s) to associate non-European states to its work, especially the United States and Canada[23]. As for political scientists and geographers, their studies remained focused on an analysis of certain border spaces in Europe. The Upper Rhine Region, or more generally the case of Swiss cross-border cooperation, was the subject of several publications in this respect[24], but in the 1990s, the first case studies of East European Euroregions were also published, such as *Die Euroregionen im Osten Deutschlands (The Euroregions in East Germany)*, published by Karl Eckhart and Hartmut Kowalke[25]. Political scientists also started to develop specific theories on cross-border governance. The pioneers in this field of study were Peter Schmitt-Egner, Bernd Gross and Gerhart Brunn in Germany and Guy Saez, Jean-Philippe Leresche and Michel Bassand in Switzerland[26].

However, in the 1990s, a shift of interest can be observed in the research on cross-border cooperation away from case-study approaches

---

Einrichtungen, Baden-Baden, Nomos, 1995; G. Gutt, *Grenzüberschreitende kommunale Zusammenarbeit nach dem Karlsruher Abk*ommen, Baden-Baden, Nomos, 1999.

22 Conseil de l'Europe, *Manuel de coopération transfrontalière à l'usage des collectivités locales et territoriales en Europe*, Strasbourg, 1996.

23 B. Wassenberg, *Histoire du Conseil de l'Europe*, Brussels, PIE Peter Lang, 2009, p. 84.

24 J. Beck, *Netzwerke in der transnationalen Regionalpolitik, Rahmenbedingungen, Funktionsweise, Folgen*, Baden-Baden, Nomos, 1997; R. Zoller Schepers, *Grenzüberschreitende Zusammenarbeit am Oberrhein*, Dissertation, St. Gallen, 1998; M. Lezzi, *Raumordnungspolitik in europäischen Grenzregionen zwischen Konkurrenz und Zusammenarbeit. Untersuchungen an der EG-Außengrenze Deutschland-Schweiz*, Zürich, Geographisches Institut, 1994; K. Becker-Marx, and C. Jentsch (eds.), *Es ist Zeit für den Oberrhein*, Mannheim, Südwestdeutsche Schriften, 1996; B. Reitel, *Une première approche des dynamiques urbaines dans le Rhin supérieur*, Strasbourg, Université Louis Pasteur, 1996; H. Müller-Schnegg, *Grenzüberschreitende Zusammenarbeit in der Bodenseeregion, Bestandsaufnahme und Einschätzung der Verflechtungen*, Herisau, Walker Project Partners, 1994; B. Speiser, "Europa am Oberrhein, der grenzüberschreitende Regionalismus am Beispiel der oberrheinischen Kooperation", in *Schriften der Regio 13*, Basel, 1993.

25 K. Eckart, and H. Kowalke (eds.), *Die Euroregionen im Osten Deutschlands*, Berlin, Schriftenreihe der Gesellschaft für Deutschlandforschung, 1997.

26 B. Gross, and P. Schmitt-Egner, *Europas kooperierende Regionen, Rahmenbedingungen und Praxis transnationaler Zusammenarbeit deutscher Grenzregionen in Europa*, Baden-Baden, Nomos, 1994; G. Brunn, and P. Schmitt-Egner, *Grenzüberschreitende Zusammenarbeit in Europa*, Baden-Baden, Nomos, 1998; G. Saez, J. P. Leresche, and M. Bassand, *Gouvernance métropolitaine et transfrontalière*, Paris, L'Harmattan, 1997; P. Le Galès, and C. Lequesne, *Les paradoxes des régions en Europe*, Paris, La Découverte, 1997.

on specific local and regional actors towards more general research on European integration[27]. It was thanks to the political scientists that cross-border relations were seen henceforth as part of – if not International Relations – at least the European integration process. In the early 1990s, political theories on European regionalism became popular, especially in the context of the reform process of the European Community that led to the adoption of the Maastricht Treaty in 1992[28]. With the creation of the Committee of the Regions and the introduction of the principle of subsidiarity in article 3B, an important role was allocated to the European Union's (EU) regions which would henceforth take part not only in the European institutions' decision-making process, but also more generally in the European integration process. This new situation led political scientists to develop a "Europe of regions" approach, in which regional and local authorities played a leading part, sometimes as far as to threaten – at least theoretically – the pertinence of the nation states. German researchers such as Rudolf Hrbek, Bullmann, Gerhard Brunn and Niels Lange[29] published books on the revival of European regionalism and it was in this context that some cross-border cooperation specialists, such as Sylvia Raich or Ingrid Grom[30], developed a new approach in which border regions had a specific function in the European integration process. As far as the implementation of the Single European Market was concerned, border regions were no longer regarded as disadvantaged entities at the periphery, but rather as the centrepiece at the heart of the process of eliminating all European economic borders. The free movement of goods, services, capital and people could be observed in the border regions[31]. Cross-border cooperation became a tool for the implementation

27 B. Wassenberg, "Introduction", in B. Wassenberg, and J. Beck (eds.), *Living and Researching Cross-Border Cooperation (Volume 3): The European Dimension*, Stuttgart, Steiner Verlag, pp. 29-51.

28 M. Keating, "The Invention of Regions: Political Restructuring and Territorial Government in Western Europe", in *Environment and Planning C*, No. 15, 1997, pp. 383-98; B. Kohler-Koch, *Interaktive Politik in Europa: Regionen im Netzwerk der Integration*, Opladen, Leske & Budrich, 1998; A. Benz *et al.*, *Regionalisierung: Theorie, Praxis, Perspektiven*, Opladen, Leske & Budrich, 1999.

29 R. Hrbek, and S. Weyand, *Betrifft: Das Europa der Regionen, Fakten, Probleme, Perspektiven*, München, Beck, 1994; U. Bullmann, *Die Politik der dritten Ebene, Regionen im Europa der Union*, Baden-Baden, Nomos, 1994; G. Brunn, *Region und Regionsbildung in Europa*, Baden-Baden, Nomos, 1996; N. Lange, *Zwischen Regionalismus und europäischer Integration*, Baden-Baden, Nomos, 1998.

30 I. Grom, *Die grenzüberschreitende Zusammenarbeit als Beitrag zur Förderung der europäischen Integration*, Berlin, Verlag Dr. Köster, 1995; S. Raich, *Grenzüberschreitende und interregionale Zusammenarbeit in einem Europa der Regionen*, Baden-Baden, Nomos, 1995.

31 C. Ricq, *Les Cantons frontaliers et l'intégration européenne, Acte Unique européen, 1992*, report of the symposium, 28th October 1988, Geneva, p. 56.

of the European integration project and border regions became either models or testing grounds for the success of this project[32].

From 2000 onwards, historians also started to include the issue of regional cooperation in their research on the history of European integration. Within the European Commission's liaison group of historians, Marie-Thérèse Bitsch organised a conference on *Le fait régional dans la construction européenne (The regional element in European integration)*, published in 2002[33]. It included several contributions on cross-border cooperation and its relevance for the European integration process[34]. Researchers in history, geography and in political sciences carried out more case studies on the development of specific cross-border regions – the Upper Rhine region[35], Saar-Lor-Lux[36] or the border regions between France and Spain[37] – where they examined in more detail the link between local and regional cooperation and European integration. The political scientists emphasised the relevance of new governance theories within the EU that included local and regional actors in the decision-making process[38]. The so-called multi-level governance approaches[39] were also applied to cross-border cooperation, as for example in the works of

32 K. H. Lambertz (ed.), *Die Grenzregionen als Labor und Motor kontinentaler Entwicklungen in Europa. Berichte und Dokumente des Europarates sowie Reden zur grenzüberschreitenden Zusammenarbeit in Europa*, Zürich, Baden-Baden, Dike/ Nomos, 2010.

33 M. T. Bitsch (ed.), *Le fait régional et la construction européenne*, Brussels, Bruylant, 2003.

34 Y. Lejeune, "Le cadre juridique de la coopération transfrontalière décentralisée franco-belge", pp. 287-301; C. Gengler, "Le Grand-Duché de Luxembourg, un micro-État à la croisée du fait régional et de la coopération transfrontalière", pp. 281-405; B. Wassenberg, "Le management interculturel des relations transfrontalières: l'exemple du Rhin supérieur", pp. 405-431; R. Hudemann, "Expériences frontalières et mémoires dans l'espace Saar-Lor-Lux", pp. 431-443.

35 M. Nagelschmidt, "Das oberrheinische Mehrebenensystem", in *Schriften der Regio*, No. 20, Basel, 2005; B. Wassenberg, *La coopération transfrontalière franco-germano-suisse dans l'espace du Rhin supérieur*, Brussels, PIE Peter Lang, 2007.

36 J. Leinen (ed.), *Saar-Lor-Lux. Eine Euro-Region mit Zukunft?*, St. Ingbert, Röhrig Universitätsverlag, 2001; B. Groβ *et al.* (eds.), *SaarLorLux von A bis Z*, Baden-Baden, Nomos, 2005.

37 J. B. Harguindéguy, *La frontière en Europe: un territoire? Coopération transfrontalière franco-espagnole*, Paris, L'Harmattan, 2007.

38 See the different contributions in J. Beck, and B. Wassenberg (eds.), *Grenzüberschreitende Zusammenarbeit leben und erforschen (Band 2): Governance in deutschen Grenzregionen*, Stuttgart, Steiner Verlag, 2011.

39 W. Grande, "Multi-Level Governance: Institutionelle Besonderheiten und Funktionsbedingungen des europäischen Mehrebenensystems", in E. Grande, and M. Jachtenfuchs (eds.), *Wie problemlösungsfähig ist die EU? Regieren im europäischen Mehrebenensystem*, Baden-Baden, 2000, p. 11.

Petra Newrly (*Transnationaler Regionalismus. Die grenzübergreifende Zusammenarbeit am Oberrhein*)[40], of Almut Kriele, Urs Lesse and Emaunel Richter (*Politisches Handeln in transnationalen Räumen*)[41]. Whereas sociologists such as Philippe Hamman or Georg Simmel examined the social effects of European integration or non-integration of the people living in border regions (for example the phenomenon of cross-border workers)[42], geographers returned to a more thorough analysis of border effects in the context of a Europe without "frontiers"[43].

Generally, from the 1970s to 2010, research on cross-border cooperation as a European phenomenon increased and diversified constantly. However, whereas a link was certainly established with the research on European integration, an international approach to the subject of cross-border cooperation was not a mainstream development within the human sciences research community[44].

A research interest in cross-border cooperation as an integral part of International Relations originated from a geopolitical approach to borders, developed by geographers in the 1970s, which has only recently been followed by political scientists, sociologists and historians.

2. The Growing Interest in an International Approach to Cross-Border Studies

In the 1970s, the new school of critical political geography introduced a different approach to the modern state, emphasising the influence of borders as a means of political, geographical, economic and social delimitation[45]. Under the leadership of Noel Parker in the UK, the notions

---

40 P. Newrly, *Transnationaler Regionalismus. Die grenzübergreifende Zusammenarbeit am Oberrhein*, Münster, Lit-Verlag, 2002.

41 A. Kriele, U. Lesse, and E. Richter, *Politisches Handeln in transnationalen Räumen*, Baden-Baden, Nomos, 2005.

42 P. Hamman, *Les travailleurs frontaliers en Europe*, Paris, L'Harmattan, 2006; G. Simmel, "Der Raum und die räumlichen Ordnungen der Gesellschaft", in M. Eigmüller, and G. Vobruba (eds.), *Grenz-Soziologie. Die politische Strukturierung des Raumes*, Wiesbaden, VS-Verlag, 2006, pp. 15-23.

43 A. Paasi, "Boundaries as Social Practice and Discourse: the Finnish-Russian Border", *Regional Studies*, Vol. 33, No. 7, 1999, pp. 669-90. Cf. for example: B. Reitel *et al.*, *Villes et frontières*, Paris, Anthropos, Economica, 2002; or G. Traband, *Effacer la frontière*, Strasbourg, La Nuée Bleue, 2008.

44 B. Wassenberg, "Historiographie de la coopération transfrontalière", *op. cit.*; B. Wassenberg, "La coopération transfrontalière comme objet de recherche sur l'histoire de l'intégration européenne", *op. cit.*

45 J. Gottmann, *The Significance of Territory*, Charlottesville, University Press of Virginia, 1973; J. R. V. Prescott, *The Geography of Frontiers and Boundaries*, London, Hutchinson, 1965.

of periphery and marginality were put forward in order to identify borders as structuring elements of a political space or sphere of influence[46]. In France, the geopolitical view on borders was particularly described by Michel Foucher, who published a book titled *Fronts et frontières* in 1988[47]. This more general geopolitical approach to borders allowed for cross-border cooperation to be regarded not as a particular European phenomenon, but as an international one which could be observed in all parts of the world where two frontiers separated two national states. At the crossroads between geography and political sciences, the mainstream research of regional area studies has also described the international dimension of frontiers, but with a different objective. The interest was less in the geopolitical function of borders than in the local governance dimension within a regional space divided by an international border. In this area of study, American authors like Oscar Martinez published as early as in 1986 the first comparative analysis, *Across Boundaries, Cross-Border Interactions in Comparative Perspective*[48]. In 1994, Martinez also published the case study *Life and Society in the US-Mexico Borderlands*[49]. The geographical scope of research was thus broadened beyond the European continent to other cross-border cooperation regions in the world.

In Europe, the first political scientist to approach cross-border cooperation from a more internationally oriented, geopolitical angle was the Scot Malcolm Anderson. Although, in 1983, he started to examine cross-border regions in Western Europe[50], he enlarged his research together with Eberhart Bort after the fall of the Berlin Wall to focus on the East European borders at the former Iron Curtain. In their book *Boundaries and Identities: the Eastern Frontier of the European Union* (1996), they mainly identified the existing gap between West and East European regions[51]. But they were also confronted with new issues of cross-border cooperation relating to the problem of stabilising borders, which led them to publish two other books, one in 2000 titled *Border Regions and Security in an Enlarged Europe* and the other in

46 N. Parker, *Geopolitics of Europe's Identity: Centers, boundaries and margins*, New York, Palgrave MacMillan, 2008.

47 M. Foucher, *Fronts et frontières: un tour du monde géopolitique*, Paris, Fayard, 1988.

48 O. J. Martinez, *Across Boundaries, Cross-Border Interactions in Comparative Perspective*, El Paso, Texas Western Press, 1986.

49 O. J. Martinez, *Border People. Life and Society in the U.S.-Mexico Borderlands*, Tucson, University of Arizona Press, 1994.

50 M. Anderson, *Frontier Regions in Western Europe*, London, Frank Cass, 1983.

51 M. Anderson, and E. Bort, *Boundaries and Identities: the Eastern Frontier of the European Union*, Edinburgh, International Social Science Institute, 1996.

2001, *The Frontiers of the European Union*[52]. The role of cross-border cooperation for security in Europe served to establish a definite link with International Relations, at least as far as the political sciences were concerned, as proves, for example, Didier Bigo's contribution on "Border Regimes, Police Cooperation and Security in an Enlarged European Union", published in *Europe Unbound: Enlarging and Reshaping the Boundaries of the European Union* in 2002[53]. In the Association of European Border Region's handbook on Euroregions, published in 2008 by Jens Gabbe and Victor von Malchus, this role of cross-border cooperation was identified as one of "promoter of democratic security"[54]. Political scientists then started to analyse cross-border regions in relation to the EU's Neighbourhood Policy from this viewpoint. Cross-border cooperation was regarded as a tool of the EU's foreign policy and thus as part of International Relations. Several publications of political scientists point in this direction, for example Sandra Lavenex's articles on "EU External Governance in Wider Europe", "A Governance Perspective on the European Neighbourhood Policy: Integration beyond Conditionality?" and (together with Frank Schimmelfennig) "EU Rules beyond EU Borders: Theorizing External Governance in European Politics" published in the *Journal of European Public Policy* (2004, 2008 and 2009)[55] or Tanya Radchuk's article on "Contested Neighbourhood, or how to Reconcile Differences" published in the *Journal of Communist Studies and Transition Politics* (2001)[56]. However, this new function of cross-border cooperation in International Relations somehow altered the original objectives of cross-border relations. Whereas they had been regarded in the 1970s in Europe as a means to overcome the notion of

[52] M. Anderson, *Border Regions and Security in an Enlarged Europe*, Florence, European University Institute, 2000; M. Anderson, and E. Bort, *The Frontiers of the European Union*, Basingtoke, Palgrave, 2001.

[53] D. Bigo, "Border Regimes, Police Cooperation and Security in an Enlarged European Union", in J. Zielonka (ed.), *Europe Unbound: Enlarging and Reshaping the Boundaries of the European Union*, London, Routledge, 2002, pp. 213-39.

[54] J. Gabbe, and V. F. von Malchus, *Zusammenarbeit europäischer Grenzregionen*, Baden-Baden, Nomos, 2008, p. 5.

[55] S. Lavenex, "EU External Governance in 'Wider Europe", in *Journal of European Public Policy*, Vol. 11, No. 4, 2004, pp. 680-700; S. Lavenex, "A Governance Perspective on the European Neighbourhood Policy: Integration beyond Conditionality?", in *Journal of European Public Policy*, Vol. 15, No. 6, 2008, pp. 938-55; S. Lavenex, and F. Schimmelfennig, "EU Rules beyond EU Borders: Theorizing External Governance in European Politics", in *Journal of European Public Policy*, Vol. 16, No. 6, 2009, pp. 791-812.

[56] T. Radchuk, "Contested Neighbourhood, or how to Reconcile the Differences", in *Journal of Communist Studies and Transition Politics*, Vol. 27, No. 1, 2011, pp. 22-49.

frontiers as "scars of history"[57], i.e. as part of a reconciliation process, or in the 1990s as part of the European integration process, they could now acquire a much more negative and almost aggressive function. Within the realm of the EU's Neighbourhood Policy, cross-border cooperation could become part of a neo-imperialist strategy of the EU's power enlargement to the East. Agreeing with new historicist theories that associate EU enlargement with reasons of territorial conquest, Jan Zielonka published for example two important works pointing in this direction: *Europe Unbound: Enlarging and Reshaping the Boundaries of the European Union* in 2002[58] and *Europe as an Empire: the Nature of the Enlarged European Union* in 2006[59]. In a recent article in the Review *International Affairs*, he went as far as to ask: "Europe as a Global Actor: Empire by Example?"[60]. In 2008, Gabriel Popescu, in his article called "The Conflicting Logics of Reterritorialisation: Geopolitics of Euroregions in Eastern Europe" which was published in the Journal *Political Geography*[61], identified, from a more geopolitical viewpoint, the role of cross-border cooperation as an element permitting the reshaping of International Relations.

For geographers, from the end of the 1990s on, the geopolitical dimension of borders and especially their re-territorialisation function was becoming more and more obvious. Indeed it was a geographer, Anssii Paasi, who first developed a constructivist approach to borders which allowed for a more sociological view on re-territorialisation processes, especially when observed at the Eastern borders of the EU[62]. In his article "Fences and Neighbours in the Postmodum World: Boundary Narratives in Political Geography" published in 1998 in the Journal *Progress in Human Geography*, this new approach to borders in International Relations was clearly apparent[63]. The geopolitical element was then further underlined

57 This is the traditional quote of Alfred Mozer for qualifying the objective of cross-border cooperation, cf. B. Wassenberg, "Qu'est-ce qui motive la coopération trans-frontalière dans l'espace du Rhin supérieur?", in B. Wassenberg (ed.), *Vivre et penser la coopération transfrontalière (Volume 1): les régions frontalière françaises*, p. 96.

58 J. Zielonka (ed.), *Europe Unbound: Enlarging and Reshaping the Boundaries of the European Union*, London, Routledge, 2002.

59 J. Zielonka, *Europe as Empire: the Nature of the Enlarged European Union*, Oxford, University Press, 2006.

60 J. Zielonka, "Europe as a Global Actor: Empire by Example?", in *International Affairs*, Vol. 84, No. 3, 2008, pp. 471-84.

61 G. Popescu, "The Conflicting Logics of Cross-Border Reterritorialization: Geopolitics of Euroregions in Eastern Europe", in *Political Geography*, Vol. 27, 2008, pp. 418-38.

62 A. Paasi, "Boundaries as Social Practice and Discourse: the Finnish-Russian Border", in *Regional Studies*, Vol. 33, No. 7, 1999, pp. 669-90.

63 A. Paasi, "Fences and Neighbours in the Postmodern World: Boundary Narratives in Political Geography", in *Progress in Human Geography*, Vol. 22, No. 2, 1998, pp. 186-207.

by his colleagues in France, whether on an international scale such as in Michel Foucher's book *L'obsession des frontières* published in 2007[64], or at the European level, as shown for example in the recent publications of Anne-Laure Amilhat-Sazy and Marie-Christine Fourny (*Après la frontier, avec la frontier: nouvelles dynamiques transfrontalières en Europe*) in 2006[65], or of Bernard Reitel (*Villes et frontières. Un jeu de construction de territories*) in 2010[66]. Last but not least, historians discovered the geopolitical functions of cross-border cooperation in their research on area studies in the history of International Relations. They examined the development of intra- or extra-European neighbouring areas, *les espaces de voisinage*, from a geopolitical perspective, in order to understand the political, economic and cultural representations between bordering regions, states or groups of states[67]. In 2010, Jean-Christophe Romer and Denis Rolland published a special edition of *Matériaux pour l'histoire de notre temps* dedicated to this approach to the history of International Relations, and cross-border cooperation was included as a specific case study[68].

Research on cross-border relations has therefore been progressively situated between studies on European integration and on International Relations. In more recent works published by authors of various scientific disciplines, this position was confirmed by a tendency towards comparative approaches. For example, in the late 1990s, James Scott led a number of comparative research projects on cross-border cooperation within Europe and in North America or between Germany and Poland[69]. From a more theoretical point of view, Joachim Blatter published a book

---

64 M. Foucher, *L'obsession des frontières*, Paris, Perrin, 2007.

65 A. L. Amilhat-Szary, and M. C. Fourny (eds.), *Après la frontière, avec la frontière: nouvelles dynamiques transfrontalières en Europe*, Paris, Editions de l'Aube, 2006.

66 B. Reitel, *Villes et frontières. Un jeu de construction de territoires*, habilitation à diriger des recherches, Bâle, Berlin, Strasbourg, Université de Paris Ouest-Nanterre-La Défense, 2010.

67 J. C. Romer, "Préface, les espaces de voisinage en Europe", in *L'approche pluridisciplinaire de la coopération transfrontalière. Les jalons pour un travail de recherche interdisciplinaire*, *op. cit.*

68 B. Wassenberg, "Le voisinage de proximité: les eurorégions "géopolitiques" aux frontières externes de l'UE (1993-2009)", in D. Rolland, and J. C. Romer, "Espaces de voisinage", *Les espaces de voisinage. De Monroe à la politique européenne de voisinage. Identité et altérité, Matériaux pour l'histoire de notre temps*, No. 97 and 98, janvier-juin 2010, pp. 45-49.

69 J. Scott, "Planning Cooperation and Transboundary Regionalism: Implementing Policies for European Border Regions in the German-Polish Context", in *Environment and Planning C*, No. 16, 1998, pp. 605-624; J. Scott, "European and North American Contexts for Cross-Border Regionalism", in *Regional Studies*, Vol. 33, No. 7, 1999, pp. 605-18; J. Scott, "The EU and 'Wider Europe': Toward an Alternative Geopolitics of Regional Cooperation?", in *Geopolitics*, Vol. 10, 2005, pp. 429-454.

called *Entgrenzung der Staatenwelt* (Debordering the world of states) in 2001[70] which compared cross-border relations between Europe and North America. In 2008, Bruno Dupeyron published an intra-European comparison between cooperation in the Upper Rhine region and in the Pyrenees[71].

Finally, the new interdisciplinary research project *Living and Experiencing Cross-Border Cooperation*, launched in 2008 by the historical research group *Frontières, acteurs et représentations d'Europe* of the University of Strasbourg and implemented in cooperation with the Euro-Institute Kehl, confirmed this intermediate position of cross-border cooperation between European integration and studies on International Relations. Whereas the first two publications on French border regions and on governance in German border regions[72] focused on the origins of cross-border cooperation in Europe and on the issue of multi-level governance in the context of European integration, the subsequent books widened the scope of analysis. Thus, volume 3 on the European dimension not only dealt with the link between border regions and the European Union, but also examined cooperation within the Council of Europe, emphasising the geopolitical role of cross-border cooperation as a stabilising factor at the external borders of the greater Europe, ranging from Turkey to Russia and beyond to the Caucasus[73]. Case studies such as the Black Sea and the Adriatic Sea Euroregions[74] illustrate how important questions are raised by cross-border cooperation, which implies not only regional actors, but also several national states situated around a sea, such as power struggles over energy resources or strategic access to water routes. Taking the example of "sensitive" border regions, Volume 4 of the research project further underlined the geopolitical aspects of border relations and their impact on International Relations[75]. For example, a

---

70 J. Blatter, *Entgrenzung der Staatenwelt? Politische Institutionenbildung in grenzüberschreitenden Regionen in Europa und Nordamerika*, Baden-Baden, Nomos, 2000.

71 B. Dupeyron, *L'Europe au défi de ses régions transfrontalières, expériences rhénane et pyrénéenne*, Bern, Peter Lang, 2008.

72 B. Wassenberg (ed.), *Vivre et penser la coopération transfrontalière (Volume I): les régions françaises*, *op. cit.*; J. Beck, and B. Wassenberg (eds.), *Grenzüberschreitende Zusammenarbeit leben und erforschen (Band 2), Governance in deutschen Grenzregionen*, *op. cit.*

73 B. Wassenberg, "Introduction", in B. Wassenberg, and J. Beck (eds.), *Living and Researching Cross-Border Cooperation (Volume 3): The European Dimension*, *op. cit.*

74 Cf. articles by G. Cordiale, "The Council of Europe Acting for Cross-Border Cooperation. The Adriatic Euroregion and the Black Sea Euroregion", pp. 181-187; and E. Rodaro, "The Adriatic Euroregion and the Black Sea Euroregion. Overview and Perspectives", pp. 187-209.

75 B. Wassenberg, and J. Beck (eds.), *Vivre et penser la coopération transfrontalière (Volume 4): les régions frontalières sensibles*, *op. cit.*

contribution on cross-border cooperation between Spain and Morocco illustrated the problems of the "Fortress Europe" in terms of immigration and security[76]. This series of publications from this research project has made apparent the growing importance of studies on the international dimension of cross-border cooperation[77]. It was put forward as one of the major challenges for future interdisciplinary work on this subject.

3. Conclusion: Cross-Border Cooperation: an International Relations Research Target

The link between cross-border cooperation and International Relations is all but evident. Research by all of the human science disciplines has so far been rather limited and furthermore mostly restricted to the study of European cross-border cooperation. Whereas from the 1970s, geographers were approaching border studies from an international, geopolitical standpoint, cross-border cooperation had been regarded as part of the reconciliation process in Europe after the Second World War: overcoming the border in order to heal the scars of history was the main objective of cross-border cooperation and also the main focus of the different researchers' analyses. Thanks to the political scientists' interest in regionalism in the 1990s, cross-border cooperation then became linked to the European integration process, but a more general connection with International Relations was only really established by researchers at the beginning of the 2000s. The leading role was again taken by the geographers with a renewed geopolitical approach that then spread to other disciplines. However, for studies in International Relations to fully include cross-border cooperation as a subject for interdisciplinary research, the role of this micro-diplomacy has still to be recognised in theories on International Relations. There is still therefore a strong need for all human science disciplines – geographers, political scientists, sociologists, legal specialists – to join their efforts in order to deepen the link between cross-border cooperation and International Relations.

76 Article by S. Latmani, "La coopération transfrontalière entre le Maroc et l'Espagne. Limites et perspectives", pp. 271-281.

77 The series is completed by volumes 5 and 6: J. Beck, and B. Wassenberg (eds.), *Grenzüberschreitende Zusammenarbeit leben und erforschen (Bd. 5): Integration und (trans-)regionale Identitäten, op. cit.*; J. Beck, and B. Wassenberg (eds.), *Vivre et penser la coopération transfrontalière (Volume 6): vers une cohésion territoriale?*, Stuttgart, Steiner Verlag, 2014.

From Europhoria to Crisis?

Cross-Border Cooperation, Euroregions and Cohesion

James W. Scott

Karelian Institute, University of Eastern Finland

1. Introduction

Cross-border co-operation (CBC) has been promoted by the European Union on the assumption that national and local identities can be complemented (perhaps partly transcended) and goals of co-development realised within a broader – a European – vision of community. As such, borders have been used as explicit symbols of European integration, political community, shared values and, hence, identity by very different actors[1]. Consequently, the Euroregion concept has proved a powerful tool with which to transport European values and objectives and to reinforce the EU's political identity[2]. From their relatively modest beginnings as local expressions of good "neighbourliness" in Dutch-German border regions, Euroregions have been appropriated and promoted by policy entrepreneurs at the local, national and European level in order to signal a new quality of interstate integration within Europe. "De-bordering" within the enlarged European Union and new cross-border relations in Central and Eastern Europe indicate that not only states but citizens, communities and regions have chosen to open new avenues of communication with their neighbours across national boundaries. Furthermore, in those

---

1 K. Lepik, "Euroregions as Mechanisms for Strengthening of Cross-Border Cooperation in the Baltic Sea Region", in *TRAMES*, Vol. 13, No. 3, 2009, pp. 265-284; M. Perkmann, "Policy Entrepreneurship and Multilevel Governance: a Comparative Study of European Cross-Border Regions", in *Environment and Planning C*, Vol. 25, No. 6, 2007, pp. 861-879; G. Popescu, "The Conflicting Logics of Cross-Border Reterritorialisation: Geopolitics of Euroregions in Eastern Europe", in *Political Geography*, Vol. 27, 2008, pp. 418-438.

2 See E. Bojar, "Euroregions in Poland", in *Tijdschrift voor economische en sociale geografie*, Vol. 87, No. 5, 2008, pp. 442-447.

contexts where states have (re)gained their independence and new borders have emerged[3], Euroregions, cross-border city partnerships and similar cooperation vehicles have also come into being[4].

Research interest in CBC (or Territorial Cooperation as it is now conceptualised in EU parlance) has been spurred by enlargement and integration contexts but also the more strategic focus of European Cohesion policy shifts that have elevated CBC to the paradigmatic status it now enjoys. By the same token, this research also reveals that the normative political language of integration often contrasts with local realities. Rather than represent a new governance ideal, cross-border co-operation often reflects competing territorial logics at the EU, national, regional and, local levels and conflicting attitudes towards more open borders[5]. One major weakness in understanding CBC as a new border-spanning form of governance has been the frequent neglect of socio-political contexts and the application of largely untested assumptions based on "new regionalist" theoretical perspectives[6]. Unsurprisingly, research has cast serious doubt upon the notion that induced and institutionally "thick" Euroregions can by themselves lead to a transcending of boundaries in policy terms. Partly as a response to limitations of the governance approach, border scholars, especially since the turn of the Millennium, have elaborated constructivist understandings of CBC as a contested regional development project[7]. Consequently, several issues of theoretical and practical interest have emerged from the state-of-the-art research on cross-border cooperation within the context of European Cohesion:

3 Such as the Baltic States, Ukraine, Moldova, Russia and the Balkans.

4 K. Lepik, *Cross-Border Cooperation Institutional Organisation*, Saarbrücken, Lampert Publishing, 2012; T. Zhurzhenko, "Regional Cooperation in the Ukrainian-Russian Borderlands: Wider Europe or Post-Soviet Integration?", in J. W. Scott (ed.), EU Enlargement, Region Building and Shifting Borders of Inclusion and Exclusion, Aldershot, Ashgate, 2006, pp. 95-111.

5 G. Popescu, "The Conflicting Logics of Cross-Border Reterritorialisation: Geopolitics of Euroregions in Eastern Europe", *op. cit.*, pp. 418-438.

6 E. Gualini, *Cross-Border Governance: Inventing Regions in a Trans-national Multi-level Polity*, DISP, 152, 2003, pp. 43-52.

7 H. J. Bürkner, "Regional Development in Times of Economic Crisis and Population Loss: The Case of Germany's Eastern Border Regionalism", in J. W. Scott (ed.), *EU Enlargement, Region-building and Schifting Borders of Inclusion and Exclusion*, *op. cit.*, pp. 207-215; O. Kramsch, and B. Hooper (eds.), Cross-Border Governance in the European Union, London and New York, Routledge, 2004; M. Perkmann, "Policy Entrepreneurship and Multilevel Governance: a Comparative Study of European Cross-Border Regions", *op. cit.*, pp. 861-879; V. Kolossov, and J. Scott, "Karelia: A Finnish-Russian Borderland on the Edge of Neighbourhood", in H. Eskelinen, I. Liikanen, and J. Scott (eds.), *On the Edge of Neighbourhood: Regional Dimensions of the EU-Russia Interface*, London and New York, Routledge, 2012, pp. 194-210.

1. Relations between material and discursive regionalism and abstract and real (everyday) spatial contexts;
2. The role of historical memory and national development in framing border-related issues;
3. The limits to "Europeanisation" through EU policies and initiatives;
4. The role of local milieu and socio-political contexts;
5. The multiple roles of different actor networks in promoting transboundary co-operation.

Hence, looking back at the evolution of CBC since 1989 (and thus with the benefit of hindsight not available earlier), we might argue that the development of Euroregions has reflected in rather clear contours central elements of the EU's crises in terms of its political identity, governance, social agendas and economic perspectives. The limits of paradigmatic thinking with regard to CBC and Euroregions have become particularly evident in the case of Central and Eastern Europe. Geopolitical contexts on the one hand, but also contradictions within the European Union on the other, have arguably contributed to a "re-nationalisation" of policy contexts for local and regional cross-border co-operation.

Given these apparently contradictory experiences, what can be said about the trajectory of CBC as an element of EU political identity and within the context of Cohesion Policy? Recently conducted research reflects tensions between realist regional policy concerns related to national development and more idealistic policy imperatives that seek to create alternative, border-transcending territorial contexts for regional policy[8]. Idealistic notions of Cohesion as exemplified by CBC (Territorial Cooperation in EU parlance) co-exist in a parallel policy level driven by geostrategic thinking (e.g. spatial planning) but that only partially interpenetrates the realist policy environment. At the level of EU Cohesion Policy, the direct coupling of CBC with regional development goals appears to be shifting towards more territorially flexible arrangements and a focus on "place-based strategies" and "integrated territorial investments" which can be potentially implemented in cross-border and transnational contexts. Nevertheless, the overall resources available for genuinely border-transcending regional development are but a small fraction of the overall EU structural funds budget, which is targeted largely at newer and poorer member states.

---

[8] Among other sources, reference is made to research funded by the EU's 7th Framework Programme for Research and Innovation (EUBORDERREGIONS, contract 266920) the ESPON TERCO project (www.esponterco.eu) and research carried out by the author within the framework of the Distinguished Visiting Scientist Programme of the Hungarian Academy of Sciences (2012-2013).

As a result, and as will be discussed in more detail below, we can detect highly differentiated patterns in terms of the appropriation of CBC as a regional development instrument. Admittedly, there has been a general shift away from prioritising CBC in Cohesion Policy. However, while CBC has become routinised and even independent of structural fund support in western European contexts, it is seen more in instrumental, even opportunistic, terms in Central and European countries.

2. Cross-Border Cooperation as a Practical and Conceptual Form of "Europeanisation"

The European Union has in large part been a project of transcending borders and national divisions and thus creating conditions for durable peace, prosperity and more effective interstate co-operation. In doing this, the development of the EU has gone hand in hand with the emergence of shared policy agendas and policy instruments that have created new links between member states. At the same time, the EU has sought to create a supranational community based on a shared sense of political, social and cultural identity. As a result, processes of Europeanisation – defined in terms of a gradual diffusion of transnational understandings of citizenship, identity and governance – are closely related to changing political understandings and uses of Europe's many state borders[9].

Since the end of the 1980s, cross-border co-operation has been firmly embedded within EU structural policies, Cohesion Policy in particular. It is also an important element of the new European Neighbourhood Programme operating at the EU's external borders. Consequently, questions of territoriality, regional policy and cross-border cooperation have been central to the emergence of the European Union as a political community. According to Manzella and Mendez[10], regional policy has been "perceived as a crucial instrument for the identity of a European model of society, and for the legitimacy and viability of the whole political process of integration". However, in order to signify something more than the sum of national concerns, the EU has needed an *exceptional and idealist* narrative that goes beyond state-centred political thinking and that is open-ended – territorially and conceptually. Arguably, a major element of the EU's political identity lies precisely in reconciling flexible idealism with more fixed realist territorial perspectives. EU Cohesion Policy has thus emerged concurrently

---

9 C. M. Radaelli, and R. Pasquier, "Conceptual Issues", in P. Graziano, and M. Vink (eds.), *Europeanisation: New Research Agendas*, Basingstoke, Palgrave Macmillan, 2006, pp. 35-45; J. W. Scott, and I. Liikanen (eds.), *European Neighbourhood through Civil Society Networks*, Abingdon, Taylor and Francis, 2011.

10 G. P. Manzella, and C. Méndez, *The Turning Points of EU Cohesion Policy*, Barca Report Working Paper, DG Regio, Brussels, 2009.

with paradigms of cross-border cooperation and notions of territory based on spatial relationships, cross-border and transnational networks and supranational geostrategies rather than exclusively on administrative and legalistic frameworks. A central aspect here is the recasting of national spaces as integral elements of an international political community; from this also emerges the attempt to create a common set of discourses in which various political and social issues can be negotiated.

2.1. CBC, Territorial Cooperation and Cohesion

Formally introduced by the Single Act of 1985, partly in response to the regional challenges posed by the southern enlargement, the notion of "Economic and Social Cohesion" has become a central unifying idea legitimizing an EU role in territorial development. Furthermore, Territorial Co-operation (TC) within the European Union is understood as a form of local and regional promotion of Cohesion that transcends state borders. In academic debate, territorial co-operation is most generally known as "cross-border co-operation" but, basically speaking, both terms have equivalent meanings. More pointedly, CBC/TC can be defined in terms of political projects carried out by private, state and, to an extent, third sector actors with the express goal of extracting benefit from joint initiatives in various economic, social, environmental and political fields. This has been associated with state-society paradigms which suggest that new forms of politically relevant action can (or must) increasingly take place "beyond the state" and beyond the seemingly inflexible territoriality of the state. Through new forms of political and economic interaction – both institutional and informal – it has been suggested that greater cost-effectiveness in public investment can be achieved, economic complementarities exploited, the scope for strategic planning widened and environmental problems more directly and effectively addressed.

The concept of CBC in the European context is not new; it began as a number of subnational political projects already in the 1950s between Dutch and German communities. However, it is the context of Post-Cold War change that has elevated CBC to the paradigmatic status it has enjoyed in EU policy. As such, CBC has been appropriated by the European Union as a unique social innovation and as part of the EU's new and progressive political identity[11]. CBC is thus an element of Europeanisation – which can be understood as a diffusion of supranational notions of post-national stateness and a hybrid, multilevel sense of governance, citizenship and identity[12].

---

11 J. W. Scott, "Europe of Regions", in R. Kitchen, and N. Thrift (eds.), *International Encyclopaedia of Human Geography*, Amsterdam, Elsevier, 2009, pp. 649-657.

12 J. W. Scott, and I. Liikanen (eds.), *European Neighbourhood through Civil Society Networks*, *op. cit.*

Through its support of CBC, the European Union has promoted a self-image of role model for intercultural dialogue and local/regional development[13].

During the 1990s and early 2000s, the EU was in fact busy promoting spaces for new forms of cross-border "multilevel governance". Within this context, border regions were explicitly understood to be important elements within European integration policies by representing potentially flexible vehicles with which to manage conflict and facilitate collective action in the management of social, economic and environmental issues[14]. In retrospect, it is clear that normative notions of Europeanisation through CBC were largely influenced by formal, structural understandings of transnational governance[15]. In prescriptive policy terms, this translated into an imperative of institutionalisation in which prospects for successful CBC were defined by the outcomes of a gradual and complex process of collaborative innovation and capacity-building at national, state and local levels. At the same time, intercultural dialogue, together with adequate strategies with which to reconcile and co-ordinate diverse interests, were seen to offer considerable promise for developing cross-border alliances between cities and their regions[16]. One further normative aspect in all of this was a highly optimistic assumption of new synergy effects and greater mutual benefits to the actors and localities that engage in CBC.

This governance focus on CBC has had a direct impact on Cohesion Policy through the emphasis of networked, territorially flexible and border-transcending relationships between stakeholders in local and

---

13 This is reflected, for example, in the EU's Green Paper on Territorial Cohesion (subtitled 'Turning Territorial Diversity into Strength') in which the need to develop strong cross-border linkages as well as more robust forms of regional and local cooperation with neighbouring states is emphasized (see Commission of the European Communities, 2008).

14 M. Bufon, and V. Markelj, "Regional Policies and Cross-Border Cooperation: New Challenges and New Development Models in Central Europe", in *Revista Romana de Geografia Politica* XII, Vol. 1, 2010, pp. 18-28; M. Perkmann, "The Rise of the Euroregion. A Bird's Eye Perspective on European Cross-Border Co-operation", published by the Department of Sociology, Lancaster University (www.comp.lancs.ac.uk/sociology), 2002.

15 See J. Blatter, "Explaining Cross-Border Cooperation: a Border-Focused and Border-External Approach", in *Journal of Borderlands Studies* XII, Vol. 1/2, 1997, pp. 151-175; J. Blatter, "From Spaces of Places' to 'Spaces of Flows'? Territorial and Functional Governance in Cross-Border Regions in Europe and North America", in *International Journal of Urban and Regional Research*, Vol. 28, No. 3, 2004, pp. 530-548.

16 M. van Geenhuizen, B. van der Knapp, and P. Nijkamp, "Trans-Border European Networks: Shifts in Corporate Strategy?", in *European Planning Studies*, Vol. 4, No. 6, 1996, pp. 671-682; M. Leibenath, E. Korcelli-Olejniczak, and R. Knippschild (eds.), *Cross-Border Governance and Sustainable Spatial Development*, Berlin, Springer, 2008.

regional development[17]. Cross-border co-operation involves attempts to exploit borderlands situations, using borders as a resource for economic and cultural exchange as well as for building political coalitions for regional development purposes[18]. One strategy pursued by the EU has been to couple cross-border cooperation with more general Cohesion policies through institution-building, generally, but not exclusively, in the form of so-called Euroregions. The main goal of Euroregions and similar organisations has been to promote a sense of "EU-Europeanness" through mutual learning and co-operative initiatives across borders and, at the same time, address specific regional economic, environmental, social and institutional problems. These associations have served as instruments of EU policy in sharing roles (not always as equal partners) with regional and national governments in the channelling of European regional development support into border regions. Euroregions were pioneered and developed as locally based co-operation initiatives in Dutch-German border regions as early as in the 1960s[19]. Since then, Euroregions have become part of complex policy networks at the European and national levels and have contributed to cross-border regional development, particularly in West European contexts. Indeed, the Dutch-German EUREGIO, a Euroregion with its own local council and close ties to German and Dutch state agencies, has served as a model of sorts for the development of border region associations within the European Union. In its different phases of development, CBC has been characterised by the adaptation of existing institutional structures to new opportunities and problems set by recent geopolitical changes. Given the long track record of cross-border cooperation in Western Europe it is not surprising that cooperation stakeholders in Central and Eastern Europe have emulated many of the institutions and projects pioneered within the EU.

In addition to local and regional forms of institutionalised CBC, *spatial planning* promotes a decidedly pan-European one. Indeed, one of the principal assumptions underlying cross-border planning exercises is that symbolism guides collective action by creating a sense of common understanding and "language". Alternative European geographies have been defined, among others, through symbolic planning concepts, the transnationalisation of space through networks and flexible regionalisation,

---

17 See, for example, the presentation on Multilevel Governance and CBC available at: http://www.cesci-net.eu/tiny_mce/uploaded/SimonaPohlova_091210.pdf.

18 G. Popescu, "The Conflicting Logics of Cross-Border Reterritorialisation: Geopolitics of Euroregions in Eastern Europe", *op. cit.*, pp. 418-438.

19 M. Perkmann, "Policy Entrepreneurship and Multilevel Governance: a Comparative Study of European Cross-Border Regions", *op. cit.*

and network-like forms of governance[20]. These initiatives are reflected in the European Spatial Development Perspective, or ESDP, first elaborated in 1999. Although not a community level policy in the sense of agriculture or regional development, ESDP, together with the ESPON programme of creating a Europe-wide planning database, has provided a policy framework of an advisory nature agreed by the European Ministers of Spatial Planning and that pursues sustainable economic development and socio-economic cohesion[21].

3. CBC – a Question of "Filling in the Gaps"?

Despite the above, there is no doubt that since the historic turn of events of 1989/1991 and the heady days of a "new European order" there has been a shift in the EU's focus on CBC. EU rhetoric about the benefits of CBC is today a far cry from the prosaic language of the 1990s. In the now archived EU Regional Policy website we can read that in fact "cross-border cooperation is essentially about filling the gaps [between national development]"[22]. Most recently, CBC has been subsumed within the more inclusive notion of Territorial Cooperation and its main aim remains to reduce the negative effects of borders as administrative, legal and physical barriers, tackle common problems and exploit untapped potential"[23]. It is clear from recent debate on European Cohesion that the EU stakes much of its political capital in more traditional instruments of redistribution that are nationally oriented even if subject to supranational guidelines. Indeed, the 2007-2013 budget of €8.7 billion for Territorial Cooperation amounts to a mere 2.5% of the total Cohesion Policy budget. Furthermore, a major overall share of Cohesion funds are targeted to Central and Eastern European countries where there appears to be less enthusiasm for CBC as a regional development resource.

Arguably, we see a strong dose of political realism creeping into European discussion of borders and cross-border cooperation. As some

20 A. Faludi, *Cohesion, Coherence, Cooperation: European Spatial Planning Coming of Age?*, London, Taylor and Francis, 2010.

21 Central to ESDP is a focus on regional urban systems, urban-rural relationships, access to development opportunity structures and a concern for a diverse natural and cultural heritage. These spatial strategies cross-cut traditional nationally-oriented development practice; in effect, nothing less than an "EU-Europeanisation" of regional and local political spaces is being attempted (see M. Fritsch, "European Territorialisation and the Eastern Neighbourhood: Spatial Development Co-operation between the EU and Russia", in *European Journal of Spatial Development*, Vol. 35, 2009, pp. 1-27).

22 http://ec.europa.eu/regional_policy/archive/cooperation/index_en.htm [last access on 4th October 2013].

23 *Ibid.*

observers have pointed out, hardening external borders[24] and an increasing EU emphasis on territoriality of political community[25] as well as more specifically *national* orientations in Cohesion Policy have relegated CBC to the status of a regional policy appendage. By the same token, the fact that CBC has been rhetorically delegated to the interstitial spaces between integrating national economies is partly understandable given the mixed results of past CBC initiatives. Although the promotion of territorial co-operation and a sense of cross-border "regionness" through common institutions has been intensive in theory, in practice institutionalisation patterns have been uneven – both in terms of governance capacities and their performance in terms of actual cooperation[26]. Despite undeniable successes, Euroregions have clearly not automatically guaranteed the establishment of new public and private sector alliances to address regional and local development issues. In less successful cases, for example, cross-border projects have merely served to enhance local budgets without stimulating true co-operation. Generally speaking, it has also been very difficult to stimulate private sector participation in cross-border regional development. European experience would also seem to indicate that, ironically, co-operation practices have maintained an administrative, technocratic and "official" character that as yet has not sufficiently encouraged citizen action and public-sector participation.

Perhaps as a result of these sobering experiences, the direct linking of CBC with EU Cohesion goals appears to be shifting towards more territorially flexible arrangements and a focus on "place-based strategies" and "integrated territorial investments" which can be potentially implemented in cross-border and transnational contexts. Partly in response to the institutional limitations of CBC organisations, the EU also introduced in 2006 the formal-legal option of the European Grouping of Territorial Cooperation. The concept of cross-border and transnational

---

24 A. Vitale, "The Contemporary EU's Notion of Territoriality and External Borders", in *European Spatial Research and Policy*, Vol. 18, No. 2, 2011, pp. 17-27.

25 M. Fritsch, "European Territorialisation and the Eastern Neighbourhood: Spatial Development Co-operation between the EU and Russia", *op. cit.*; J. W Scott, "Reflections on EU Geopolitics: Consolidation, Neighbourhood and Civil Society in the Reordering of European Space", in *Geopolitics*, Vol. 16, No. 1, 2011, pp. 146-175.

26 Early critical observations of cross-border co-operation are provided, for example, in European Parliament, 1997; J. Mønnesland, "Cross-Border Programmes versus Regional Development Programmes. The population Density Influence", paper presented at the *Regional Studies Association Conference*, Bilbao (Spain), 21st-24th September 1999; Notre Europe – Groupement d'Études et de Recherches, *Is the New Europe Inventing Itself in its Margins? Cross-Border and Transnational Cooperation*, 2001 (available through www. notre-europe.assoc.fr); as well as in evaluations of EU structural policies such as INTERREG (http://europa.eu.int/comm/regional_policy/sources/docoffic/official/reports/p3226_en.htm).

cooperation also remains an important aspect of European spatial planning which continues to influence the definition of Cohesion Policy's core tenets, such as: sustainable economic development, regional equity, urban-rural integration and polynucleated (and thus balanced) urban development. Although not a community level policy, spatial planning is a framework for structural transnational co-operation within the EU based on macro and mesoregionalisation processes, multilevel governance partnerships and agenda-setting in spatial development issues.

CBC thus remains an important EU principle but, in policy terms, it has become an opportunity structure that is much more targeted at regions with a high degree of institutional capacity for formal CBC cooperation on the one hand, and peripheral border regions on the other. As the ESPON TERCO project has indicated, the distribution of functioning EGTCs remains heavily weighted in favour of western European countries and functional urban regions in "Core Europe"[27]. Additionally, the overall resources available for CBC make up a small fraction of the overall EU structural funds budget and in the new programming period for 2014 will be targeted largely at newer and "poorer" member states.

4. CBC as Differential Europeanisation

CBC is an area where Europeanisation has exerted considerable adaptational pressure in countries such as Hungary, Poland, Slovakia and the Czech Republic, resulting, for example, in a flurry of cross-border regional development initiatives and Euroregions. In preparing Central and East European countries for membership, the EU adopted a strategy based on institutionalised CBC and aimed at a gradual lessening of the barrier function of national borders. These policies have also been aimed at integrating previously divided border regions in order to build a more cohesive European space. Nevertheless, Gabriel Popescu[28] has argued that this normative political language of Europeanisation (e.g. as a process of de-bordering regional development) often contrasts with local realities where cross-border co-operation reflects competing territorial logics at the EU, national, regional and local levels and conflicting attitudes towards more open borders. As a result, cross-border co-operation is not uncontested. A partial resurgence of national rivalries and historical animosities has taken place in several EU member states and has, for example, affected local co-operation between Hungary, Slovakia and Romania. Different regional interpretations of CBC thus indicate a highly

27 The ESPON-TERCO final report is available at: http://www.esponterco.eu/en/news, terco-final-report.

28 G. Popescu, *Bordering and Ordering the Twenty-first Century: Understanding Borders*, Lathham (MD), Rowman and Littlefield, 2011.

variable appropriation of Europeanisation policies. Concretely, there is a notable East-West divide in the acceptance and adaptation of CBC as a set of regional development practices.

At the same time, since 2004 the situation has changed markedly. Rather than reflecting conformism to "Core Europe", CECs (e.g. Hungary, Poland, Slovakia) now strive to actively construct Europe through a redoubled focus on national development priorities. Batory[29] in fact argues that Hungarian political interpretations of European integration (and Europeanisation) have become rather formal and statist whereas Hungary's political identity with regard to its neighbours is seen in affective terms, i.e. ethnic, cultural and historic, rather than European. In the case of the Hungarian national development strategies of 1998 and 2007 we, in fact, see a marked change. While in the first document Hungarian border areas and regional development issues related to cooperation with neighbouring states received generous coverage, the 2007 National Development Strategy[30] only gives a very brief mention of CBC. In practical terms, CBC remains as a minimalist exercise – national strategic plans generally take it into consideration as an extension of national development.

The minimalist, instrumental approach of CECs indicates a relative lack of policy mainstreaming as well as a focus on national consolidation. Reasons for difficult CBC contexts in CECs also include: lack of local capacity to promote co-operation, cumbersome EU regulations and project management rules, interstate tensions and ethnolinguistic conflict, as well as local orientations to national centres and European core regions rather than to neighbouring states[31]. Furthermore, as Popescu[32] has suggested, EU inspired strategies of institutionalised CBC in Central and Eastern Europe – an area of complex social, economic and political diversity – have tended to be

[29] A. Batory, "Kin-state Identity in the European Context: Citizenship, Nationalism and Constitutionalism in Hungary", in *Nations and Nationalism*, Vol. 16, No. 1, 2010, pp. 31-48.

[30] National Development Agency, *New Hungary Development Plan. National Strategic Reference Framework 2007-2013*, Budapest, NDA, 2007.

[31] See, for example, B. Baranyi, *Magyar-ukrán határrégió*, Debrecen, MTA Regionális Kutatások Központja, 2008; Z. Hajdú *et al.*, *Local Dimensions of a Wider European Neighbourhood: Cross-Border Relations and Civil Society in the Hungarian-Ukrainian Border Area*, Pécs, Centre for Regional Studies, Discussion Paper No. 71, 2009; T. Hárdi, "Határ menti térségek, határon átnyúló kapcsolatok", in V. Szirmai (ed.), *Közép-Dunántúl*, Pécs-Budapest, MTA Regionális Kutatások Központja – Dialóg Campus Kiadó, 2010; I. Mezei, *Városok Szlovákiában és a magyar határ mentén*, Somorja (Slovakia) and Pécs, MTA RKK (Pécs Fórum Kisebbségkutató Intézet), 2008.

[32] G. Popescu, "Geopolitics of Scale and Cross-Border Cooperation in Eastern Europe: The Case of the Romanian-Ukrainian-Moldovan Borderlands", in J. Scott (ed.), *EU Enlargement, Region Building and Shifting Borders of Inclusion and Exclusion*, Aldershot, Ashgate, 2006, pp. 35-51; G. Popescu, "The Conflicting Logics of Cross-Border Reterritorialisation: Geopolitics of Euroregions in Eastern Europe", *op. cit.*

co-opted by specific nationally defined interests: Euroregions emerging in Central and Eastern Europe are generally top-down creations, inhibiting processes of region-building through local initiative. At the same time, institutional legacies, such as strong central control, have contributed to variegated Europeanisation processes[33]. Hence, CBC in the Carpathian basin (which involves Hungary, Romania and Slovakia) is not a self-evident phenomenon and appears to have lost momentum since the days of PHARE-INTERREG. In the case of Hungary, CBC is certainly understood in terms of European Cohesion but is heavily influenced by overlying political goals of "nation-building" and improving the living conditions of ethnic Hungarian communities in neighbouring states. However, this also engenders the distrust of Hungary's neighbours who at times have interpreted CBC as a means to extend Hungarian extraterritorial sovereignty claims.

In stark contrast to these developments in CECs, CBC in Western Europe is no longer as dependent on external funding as it once was. Here, we see a routinisation of local and regional cross-border cooperation that is generally embedded within multilevel governance structures. It is therefore no coincidence, as the ESPON-TERCO project has shown, that EGTCs are concentrated in western areas of the EU. In the most "successful" – that is, the most well-organised – border regions (e.g. the Dutch-German Euroregions), public-sector and NGO co-operation has been productive in many areas, especially in questions of environmental protection, local services and cultural activities. Additionally, successful cases (e.g. German-Dutch) seem to involve a process of pragmatic incrementalism, with "learning-by-doing" procedures and a gradual process of institutionalisation. As working relationships have solidified, experience in joint project development has accumulated and expertise in promoting regional interests increased, as has the capacity of regional actors to take on large-scale problems and projects.

5. Conclusions

Transcending boundaries – at least rhetorically – remains a 'leitmotif' of European Union policies. At the same time, however, it is more than evident that processes of Europeanisation, understood in the specific terms of de-bordering and engaging in CBC, are not straightforward. Local adaptation rather than direct convergence to European norms is the rule, not the exception[34]. Furthermore, the development of a European

[33] I. Kovács Pálné, "Europeanisation of Territorial Governance in Three Eastern/Central European Countries", in *Halduskultuur*, Vol. 10, 2009, pp. 40-57.

[34] M. Brusis, "The Instrumental Use of European Union Conditionality: Regionalisation in the Czech Republic and Slovakia", in *East European Politics & Societies*, Vol. 19,

political community is also often subject to entrenched national political interests as well as a lack of agreement as to what constitutes "common" European values[35]. In addition, CBC has become clearly subordinated to the core national goals of Cohesion. There is, furthermore, a notable East-West divide in the acceptance of CBC as a set of regional development practices.

More than two decades after the end of the Cold War, borders themselves often remain an obstacle to greater political and social interaction, even if the defensive character of European borders has virtually disappeared. Reflections on CBC within the EU thus must take into consideration the longue durée nature of creating cross-border political practices at the local and regional level. Indeed, CBC has rarely produced rapid results in terms of economic growth and regional development. Furthermore, local and regional actors develop cooperation mechanisms situationally and in ways that reflect both political opportunities and social and structural constraints. Despite all the shortcomings of the EU's model of institutionalised CBC, institutional change elicited by EU policies and funding mechanisms has led to a degree of Europeanisation of co-operation contexts and thus of spatial planning and development dialogue. This is evident in the discourses, agendas and practices of cross-border actors; they very often legitimise their activities by referring to the wider political, economic and spatial contexts within which their own region must develop. Nevertheless, actual patterns of CBC practices indicate a rather disjointed and complex reality. The European Union itself cannot provide a central template for de-bordering Europe. This will rather depend on how a post-national Europe is interpreted, negotiated and constructed "at the margins".

No. 2, 2005, pp. 291-316.

35 D. Augenstein, "Religious Diversity and National Constitutional Traditions in Europe", in C. Ungureanu, and L. Zucca (eds.), *Law, State and Religion in the New Europe: Debates and Dilemmas*, Cambridge (UK), Cambridge University Press, 2012, pp. 261-280.

The Emergence and Governance of Euroregions

The Case of the EUREGIO on the Dutch-German Border

Markus PERKMANN

Business School, Imperial College London

The EUREGIO is a Dutch-German Euroregion which can be defined as a cross-border region (CBR) bringing together contiguous local authorities across European borders. With a history dating back to the 1950s, the EUREGIO is among the oldest Euroregions. The EUREGIO has also been one of the main protagonists behind the Association of European Border Regions (AEBR) that over the last decades acted to spread the Euroregion model across the European territory.

This chapter has several objectives. Firstly, it introduces the case of the EUREGIO and presents evidence on its history, organisational set-up and policies[1]. Secondly, it analyses the context conditions under which the EUREGIO emerged and the governance structures that were created as a result. Particular emphasis is put on the position and role of the EUREGIO within the wider context of the European multi-level governance framework generated by EU cohesion policy. The chapter concludes with an attempt to evaluate the success and impact of the EUREGIO and a discussion of the challenges related to the EUREGIO's

[1] The empirical evidence on the EUREGIO is derived from the author's fieldwork (M. Perkmann, "Construction of New Territorial Scales: a Framework and Case Study of the EUREGIO Cross-Border Region", in *Regional Studies*, Vol. 41, No. 2, 2007, pp. 253-266; M. Perkmann, and A. Spicer, "Healing the Scars of History: Projects, Skills and Field Strategies in Institutional Entrepreneurship", in *Organization Studies*, Vol. 28, No. 7, 2007, pp. 1101-1122; archival documentation and secondary literature (M. Schack, *Grenzüberschreitende Zusammenarbeit an der deutsch-niederländischen Grenze*, Aabenraa, Institut for grænseregionsforskning, 1998; A. Aykaç, *Transborder Regionalisation: an Analysis of Transborder Cooperation Structures in Western Europe within the Context of European Integration and Decentralisation towards Regional and Local Governments*, Sindelfingen, Libertas, 1994).

dual role as representative of local authority interests and EU cohesion policy implementation agency.

1. A Short History of the EUREGIO

The EUREGIO includes part of the Gelderland, Overijessel and Drenthe provinces on the Dutch side, and parts of the *Länder* North-Rhine Westphalia (NRW) and Lower Saxony on the German side. Comprising 13,000 km^2, the EUREGIO area has a population of approx. 3.4 m. The largest urban centres are Enschede (NL), Münster and Osnabrück (D).

The origins of the EUREGIO date back to 1958 when local authorities on both sides of the Dutch-German border organised the first cross-border conference. This event had been preceded by attempts of neighbouring local authorities to collaborate. An inter-municipal association on the German side (Interessensgemeinschaft Rhein-Ems) had been founded in 1954 by local politicians representing local districts. Among its objectives were to address the restructuring problems in the textile sector and improve the local and regional infrastructures. In 1962, this more informal association was replaced by a public law inter-municipal body (Kommunalgemeinschaft Weser-Ems). A similar process of intermunicipal integration occurred in the adjoining Dutch border area where the municipalities in the areas of Twente and Gelderland founded an association[2] with the objective of co-operating more closely with the German associations. This example was followed by a second association on the Dutch side (Samenwerkningsverband Oost-Gelderland, Regio Achterhoek). Today, these three bodies together form the EUREGIO.

The inter-municipal bodies subsequently became the major agents driving cross-border co-operation and the formation of the EUREGIO. The emerging organisation was mainly pre-occupied with pressing issues such as economic marginalisation and substandard infrastructure. In order to animate these fledgling cross-border coalitions, the partners mobilised resources via a variety of channels, initially by approaching higher level authorities such as the German state authorities, central governments and supranational agencies that were emerging at this time in Europe. The collaboration was facilitated by its administrative-institutional context. Particularly, in the German system, the two-level structure of local authorities – consisting of the municipalities on the one hand and district-type aggregations of municipalities (*Kreise*) on the other – facilitates collective action among municipalities.

2 Belangengemeemschap Twente-Gelderland, later Samenwerkningsverband Twente, today Regio Twente.

The EUREGIO 'Work Group' (today: Board) was founded in 1966 to operate as the informal board of the cross-border region. On the basis of regular meetings, it attempted to shift the EUREGIO's work from purely project-based contacts towards a programmatic collaboration. The Alfred Mozer Commission, a body responsible for cultural cooperation was founded in 1970, and simultaneously a secretariat was established comprising two units on each side of the border and funded by membership fees. Two studies, in the fields of culture and economic affairs respectively, gave the secretariats a programmatic basis for the further development of the EUREGIO.

In the mid-seventies, the Work Group was given a formal basis by means of a statute, and a common action programme was developed. This institutionalisation process culminated in the EUREGIO Council in 1978, the first cross-border regional parliamentary assembly in Europe constituted by the political delegates of the member authorities.

In 1985, the separate secretariats were merged into a single secretariat, located in the border town of Gronau (DE), employing both Dutch and German staff. On the programmatic side, a 'regional cross-border action programme' was presented in 1987, outlining the general strategy for the EUREGIO for a twenty-year period (NEI nd). The programme had been initiated via an agreement in 1984 between the German Federal ministry of economics, the Dutch ministry of economics and the Länder NRW and Lower Saxony (Niedersachsen). Funding was also provided by the European Commission. A Steering Committee was established, involving the partners in this agreement as well as the provinces, districts (*Bezirksregierungen*) and the EUREGIO. The action programme contained an economic and social analysis of the programme areas, and listed a series of measures aimed at promoting their socio-economic development.

This action programme constituted the main input for the first Operational Programme for the period 1989-1992, funded as pilot project under art. 10 ERDF[3]. When the European Commission in 1990 launched a specific programme to promote cross-border co-operation, Interreg I, the EUREGIO reacted with the speedy elaboration of a new Operational Programme based on its accumulated experience. With the advent of Interreg II (1994-1999) and Interreg III (2001-2008), the EUREGIO had established itself as strong candidate for dealing with all cross-border co-operation matters in the area.

3 European Regional Development Fund, the major component of EU cohesion policy.

2. Organisational Structure and Policies

The EUREGIO is a private-law association according to German law with approx. 130 member local authorities from both Germany and the Netherlands. A statute defines the tasks and responsibilities transferred to the EUREGIO by the inter-municipal associations and defines its bodies and their duties. The current set-up reflects the failure in the 1990s to conceive a public-law bracket for the EUREGIO as previously envisaged by its decision-makers, due to fiscal problems related to Dutch personnel to be employed by a prospective public-law body.

The objective of the EUREGIO is to promote, support and co-ordinate cross-border co-operation among its members. To this purpose, it is expected to initiate activities, design appropriate projects and raise funds. It has also been given an advisory function for citizens, organisations and authorities in cross-border matters and acts as a cross-border broker between various types of organisations and authorities. Its substantial focus is broadly defined and includes a wide variety of policy fields, from economic development measures, infrastructure, to health and education to municipal services.

The EUREGIO bodies comprise the EUREGIO Council, the Board, a secretariat and a number of 'working circles' in various policy fields.

The EUREGIO Council is a para-parliamentary body with no formal legislative competencies, and is constituted by 82 elected politicians appointed by the member authorities. In the Council, formally the most authoritative among the EUREGIO bodies, issues of general political importance are discussed. The Board is a supervisory board with approx. 20 members, and is composed of elected senior civil servants and political office-bearers of the member authorities. The 'working circles', which are flexibly composed by various expert civil servants from local and supralocal authorities and interest groups, deal with most of the project-oriented work in various policy fields, such as economic policy, social matters, the environment, culture and tourism, or 'daily border problems'. In the cultural field, the 'Mozer-Kommission', named after a former EC director general, plays a special role among the several working circles and has its own budget.

The work of the EUREGIO relies substantially on the EU regional policy support programme Interreg, specifically 'Interreg A'[4]. The procedural details relating to Interreg implementation are agreed between

[4] E. K. Prokkola, "Cross-Border Regionalization, the INTERREG III A Initiative, and Local Cooperation at the Finnish-Swedish border", in *Environment and Planning-Part A*, Vol. 43, No. 5, 2011, pp. 1190.

the partners via so-called Interreg agreements[5]. The EUREGIO is given responsibility for

- the general programme and financial management, including the elaboration of the Operational Programme,
- the collection, evaluation and preliminary approval/rejection of project applications and their submission to the SC,
- the co-ordination of the match-funds.

Project proposals are collected by the secretariat where a first pre-selection is made based on the basis of the EUREGIO's own criteria, the general Interreg requirements and the prevailing legislation for the territories of the member authorities. The EUREGIO criteria require all projects to have a genuine cross-border character in substantial, organisational, personal and financial respects.

The approved proposals are submitted to the Steering Committee, which is the highest instance in Interreg decision-making. Apart from the EUREGIO representatives, the Committee is composed of senior civil servants representing the following authorities: the Dutch ministry for Economics, several Dutch provinces, and German Land-level ministries. The SC is responsible for final project approval and requires unanimous decision-making whereby each of the authorities mentioned has one vote. The NRW *Land* administration has accepted the full and sole responsibility vis-à-vis the European Commission for the lawful implementation of the Interreg OP on behalf of the Dutch central state and Lower Saxony.

As for the substantive content for the Interreg measures, the Interreg agreements generally define six policy priority areas. Within this general framework, the EUREGIO autonomously determine the priorities within its own Operational Programme. For the period 2001-2008, it controlled approx. EUR 49 m. Interreg-related funds which are complemented by national, regional and local funds to form a total budget of EUR 111 m. This amounts to an annual expenditure of approx. EUR 4.3 per capita. The largest share of funds was dedicated to the areas of economic policy, in particular technology and innovation policy. See table three for the proportion of funding allocated to various priorities.

---

5 These agreements were concluded for each Interreg period between the involved higher level Dutch authorities (central state and provinces), the German *Länder* and the four Dutch-German Euroregions.

Table 1. EUREGIO Interreg III: Financial Summary

	EUR m
Economic policy, technology and innovation	24.5
Labour market and vocational training	8.0
Social networks, border problems, culture, education	6.4
Spatial structure	4.6
Environment, nature and agriculture	1.7
Technical assistance	3.3
Total	48.7

Figures indicate EU contributions only.

Source: EUREGIO.

In addition to the Interreg-funded activities, the EUREGIO has attracted funds from other EU and national sources. By diversifying its portfolio of activities and funding sources, the EUREGIO organisation sought to avoid overdependence on Interreg. The secretariat operates a range of activities that strengthen its profile as cross-border regional advice and citizen's service centre.

3. Decision-making and Implementation Structures: the Governance of the EUREGIO

Cross-border regions represent a specific challenge within public sector governance due to their atypical, non-nested territorial set-up: As their constituent parts – municipalities, districts and other subnational jurisdictions – belong to different nation states, they do not operate in a conventional context of public administration defined by legal competencies and decision-making mechanisms rooted in public law. Though they might be interpreted as incipient new 'scales' of territorial organisation[6], CBRs do not have 'governments' but rely on voluntary co-operation within a context of a pragmatically defined and mutually recognised set of rules. For a long time, they have operated outside public law and resorted to civil law arrangements[7] and in view of the difficulties this is not expected to change. They can therefore be seen as exemplary for a type of public

6 M. Perkmann, "Construction of New Territorial Scales: a Framework and Case Study of the EUREGIO Cross-Border Region", *op. cit.*; C. M. Johnson, "Cross-Border Regions and Territorial Restructuring in Central Europe. Room for more Transboundary SPACE", in *European Urban and Regional Studies*, Vol. 16, No. 2, 2009, pp. 177-191.

7 U. Beyerlin, "Neue rechtliche Entwicklungen der regionalen und lokalen grenzüberschreitenden Zusammenarbeit", in G. Brunn, and P. Schmitt-Egner (eds.), *Grenzüberschreitende Zusammenarbeit in Europa: Theorie – Empirie – Praxis*, Baden-Baden, Nomos, 1998, pp. 118-134.

agency that has been described as 'governance' by authors in regional and urban studies, political science and organisational studies[8]. In various forms, the principle of gover*nance* – often contrasted to govern*ment* – has widely been connected to the operation of networks among parties willing to co-operate in the absence of a hierarchical instance[9].

For these reasons, the EUREGIO's decision-making processes are more complex than suggested by its organisational set-up. As an organisation, the EUREGIO operates within a context of network relationships both on the local and supralocal level that is typical for many European CBRs. These two network realms can be characterised as 'horizontal' and 'vertical' networks, respectively. While the horizontal network concerns the collaboration and co-ordination with the local member authorities and project grant holders, the vertical network is related to the governance and monitoring of Interreg implementation where higher level authorities (i.e. federal government, regional government) and the European Commission play an important role.

In the EUREGIO, networked governance can therefore be discussed in relation to three levels: (a) the EUREGIO secretariat; (b) a horizontal border-crossing network of local authorities; (c) a vertical, 'multi-level' network of authorities including local, regional, Land, central state and EU actors.

3.1. The EUREGIO as Organisation

From an organisational viewpoint, the nodal point of the EUREGIO is the secretariat, which over the last forty years has developed into an organisation with a high degree of specialist competence. The secretariat has no constitutionally instituted competencies nor does it have any guaranteed income streams. At the same time, its mandate is relatively undefined. This enables the secretariat to act in an entrepreneurial fashion as long as it has the backing of the member authorities. This relative discretion in defining and expanding its tasks has been widely used by the EUREGIO secretariat.

This is reflected in its success in generating a stable resource flow to maintain its operations. Unlike other Euroregions, the secretariat always sought to avoid overdependence on the mostly temporary resources provided by non-local authorities. As a result, the EUREGIO

8 B. Jessop, "The Rise of Governance and the Risks of Failure: the Case of Economic Development", in *International Social Science Journal*, Vol. 50, No. 155, 1998, pp. 29-44.

9 D. Messner, *The Network Society: Economic Development and International Competitiveness as Problems of Social Governance*, London, F. Cass, German Development Institute Berlin, 1997.

disposes over a considerable income from sources not related to Interreg, notably from a membership fee charged to the member authorities. The secretariat proved successful in raising project-related funding long before Interreg was launched. The development of a strong organisational basis was crucial for providing the local actors with access to Interreg implementation. The result was that when Interreg was launched, with 15 staff members, the EUREGIO secretariat was the natural candidate for programme management. It had positioned itself as the strategy unit responsible for a range of tasks no other organisation could deal with, becoming the undisputed agency for 'mobilising the region'.

Based on its expertise and local connectedness, the secretariat exerts considerable informal influence upon EU programme implementation, for several reasons: First, by acting as a project animator, it ensures that all available funds are effectively allocated. Secondly, it has made itself indispensable as a network broker. For genuine cross-border projects, project applicants need partners on the other side of the border; the relevant contacts are usually established by the secretariat. Thirdly, the secretariat uses its administrative expertise to turn initial ideas into project applications ready for submission to the Interreg Steering Committee.

Concluding, the role of the secretariat doubles as both a network broker, securing ongoing co-operation among the members of the network, and a policy entrepreneur, continuously seeking to introduce policy innovations to widen the role and impact of the EUREGIO[10]. It developed the EUREGIO from a loose network with strong ceremonial elements into a regional policy agency that has assumed the self-defined mandate of transforming the cross-border region into a central location in North-western Europe with 20 m. consumers within 150 km. The secretariat has assumed the role of a 'functional government'[11] not on the basis of a strong formal mandate but through a skilful moderation and solidification of the horizontal and vertical policy network structures the EUREGIO is embedded in. These are discussed in the following.

3.2. The Local Network

Though common rules and obligations have been defined via formalised agreements, as a non-mandatory body, the EUREGIO relies on voluntary co-operation among local authorities. This is co-ordinated via a number of networks across the member authorities, some of which

---

10 M. Perkmann, "Policy Entrepreneurship and Multi-level Governance: a Comparative Study of European Cross-Border Regions", in *Environment and Planning C: Government and Policy*, Vol. 25, No. 6, 2007, pp. 861-879.

11 J. Beck, *Netzwerke in der transnationalen Regionalpolitik: Rahmenbedingungen, Funktionsweise, Folgen*, Baden-Baden, Nomos 1997.

have visible, formalised front-ends. The most important among these is the EUREGIO Board that acts as a supervisory committee and involves senior civil servants and political office-bearers from the most powerful and active member authorities on both sides of the border. In addition, the 'Working Circles' bring together expert civil servants from local and supralocal authorities.

The most important feature of these relationships is their *technocratic* nature as the work is carried out by expert civil servants. Such technocratic networks are usually more stable than purely 'topocratic' networks composed of politicians and tend to be committed to problem-solving as opposed to pure bargaining[12]. Decision-making and co-ordination even in technocratic networks can be notoriously difficult, especially when it comes to issues of distribution or even re-distribution[13]. In the EUREGIO, conflicts could for instance arise over the geographical distribution of project funding or decisions favouring larger authorities over smaller ones, or urban municipalities over rural ones.

The EUREGIO developed a variety of mechanisms to reduce the conflictuality and complexity of decision-making within the local cross-border networks. With the secretariat playing an active part as a network broker, the EUREGIO developed a set of norms that maintain a usually productive problem-solving climate, for instance a commitment to the 'long-term' and a sense of distributional fairness. A balanced distribution of project funding over time (to mitigate limited resources) and across space (to secure commitment) is seen as crucial for unifying particularist interests behind a common objective. There is no explicit or implicit formula for distributing Interreg-support between the two countries or between different constituencies. The secretariat enjoys a strong reputation for allocating funds even-handedly. This is why decision-making on projects is usually non-conflictual and smaller municipalities are discouraged from defecting. A further factor in reducing the complexity of decision-making in networks is a relatively small number of participants[14]. In this respect, the municipal associations in the EUREGIO network play a key role by restricting the number of (relevant) participants, while representing the smaller members. At the same time, they are important in securing the commitment of the participating municipalities, even by sometimes paying for their membership. All these structural and processual features

12 R. A. W. Rhodes, "Policy Networks: a British Perspective", in *Journal of Theoretical Politics*, Vol. 2, 1990, pp. 293-317.

13 J. Beck, *Netzwerke in der transnationalen Regionalpolitik: Rahmenbedingungen, Funktionsweise, Folgen, op. cit.*

14 D. Messner, *The Network Society: Economic Development and International Competitiveness as Problems of Social Governance, op. cit.*

of the EUREGIO taken together ensure the relative stability and coherence of this networked governance arrangement over time.

3.3. The Vertical Network

The vertical, multi-level network involving regional, national and European policy agencies in EUREGIO activities constitute an example of an EU cohesion policy implementation network[15]. The supralocal network linkages provide access to a significant amount of resources, supplementing the local resources which are complementary at best. These relationships were established long before the EUREGIO was given its mandate in Interreg implementation. Its early successes all crucially depended on the support of non-local authorities, in particular the Land NRW, the Dutch government and the European Commission.

The Dutch central government and North Rhine-Westphalia have traditionally adopted a 'CBC-friendly' attitude towards their border municipalities. The European Commission had also joined the vertical network as an important catalyst and player as early as in the 1970s, long before Interreg was launched. As a result of these developments, the EUREGIO has become an important implementation partner for the European Commission. Due to its expertise and authority in the local context, it enjoys a status that goes beyond its formal position in the vertical network. While formally the Steering Committee is the ultimate instance in the decision-making process on the allocation of Interreg funds, in practice decisions are rarely imposed on the EUREGIO actors, effectively valuing the professionalism of the secretariat.

But, apart from such motives created by the social embeddedness of network contacts, for the Dutch and NRW authorities, there is also an administrative rationale for granting relative autonomy to the EUREGIO. A senior NRW official observed: '… you can't pull projects like a rabbit out of a hat'. As the need to secure the cross-border character of Interreg projects is taken seriously, the EUREGIO has a crucial role as a project animator to secure the successful allocation of Interreg funds.

The involvement of the EUREGIO in this vertical network mirrors the general networked set-up of policy implementation in EU regional policy described as multi-level governance[16]. Given the importance of Interreg, the vertical network is also relevant for maintaining the stability of the local

---

15 H. Heinelt, and R. Smith (eds.), *Policy networks and European Structural Funds*, Aldershot, Avebury, 1996.

16 L. Hooghe, and G. Marks, "Unraveling the Central State, but How? Types of Multi-level Governance", in *American Political Science Review*, Vol. 97, No. 2, 2003, pp. 233-243; S. Piattoni, *The Theory of Multi-level Governance: Conceptual, Empirical, and Normative Challenges*, Oxford, Oxford University Press, 2010.

network described above. In fact, the local EUREGIO network operates in the 'shadow of hierarchy' cast by the Steering Committee. To avoid rejection of project proposals by the *Land* or central government, the EUREGIO makes sure that projects have the structural economic impact required by the cohesion policy regulations. The relatively clear criteria for evaluating project proposals simplify the decision-making process by providing rules that can be routinely followed. In case of doubt, the uncertainty-reducing function of these rules is complemented by the 'shadow' of authority exerted by the Dutch and NRW governments and the EU. Hence, the EUREGIO secretariat performs an interface function by mediating between the network of EUREGIO members and the supranational network crystallised in the Steering Committee. On one hand, the supranational authorities trust in the secretariat's administrative expertise and its ability to prioritise the 'objective' quality of projects over the particularist interests of project applicants. On the other, the local authorities perceive the secretariat as their agent and representative vis-à-vis the higher-level authorities.

Concluding, the described three-tiered governance structure developed by the EUREGIO provides an arrangement for co-ordinating public agency across national borders that is as effective as it can be under the existing administrative and legal limitations.

4. Conclusion and Evaluation

The EUREGIO is a cross-border region driven by local authorities and co-ordinated by a relatively independent and resourceful secretariat. This arrangement has proved a success recipe for organisational and policy growth over the last decades. In this concluding section, I address three questions that are of interest for the overall evaluation of the EUREGIO and European CBC in general. First, how should we judge the success or otherwise of Euroregions, and the EUREGIO specifically? I propose a number of criteria that can be used for assessing the relative success of CBC initiatives. Second, have the EUREGIO, and Euroregions in general, generated innovative forms of governance? In this respect, the EUREGIO has indeed had a major impact on the creation of the Euroregion model as a whole; however, to a large extent, this process was also strongly shaped by EU cohesion policy that provided opportunities for the emergence of agencies such as the EUREGIO. Third, what can be said about the impact of the EUREGIO on stimulating local cross-border integration?

First, there is considerable variation among Euroregions[17]. For many, Interreg is by far the most important source of funding, raising the

---

[17] M. Perkmann, "Cross-Border Regions in Europe. Significance and Drivers of Regional Cross-Border Co-operation", in *European Urban and Regional Studies*, Vol. 10, No. 2, 2003, pp. 153-171.

question whether these initiatives only exist *because* this type of resource is available to them. In that case, they would qualify as hardly more than 'grant coalitions'[18] that disintegrate once the funding stream runs out. On the other hand, observers have also pointed out that a certain degree of 'entrepreneurial' behaviour can indicate an effective empowerment of the regions against their central-state authorities within the context of EU integration[19]. In this sense, therefore, their ability to mobilise funding could be interpreted as success.

There is therefore a continuum between what can be called 'permanent co-operation', on the one hand, and 'instrumental co-operation', on the other, whereby the latter refers to co-operation arrangements that exist purely for the sake of taking advantage of Interreg funds. Permanent co-operation, by contrast, would apply to Euroregions that have managed to avoid total dependence on Interreg. The EUREGIO would fall into the latter group.

In fact, the EUREGIO and other more developed cases can be seen as successful instances of *policy entrepreneurship* that managed to establish themselves as trusted implementation partners for the European Commission and central state authorities. Policy entrepreneurs are actors who initiate activities within specific policy areas by taking advantage of windows of opportunity opened up within their policy environment[20]. They do this to increase the influence of their organisation or organisational unit which is often correlated to their resource basis. There are three criteria that can be used for judging whether a Euroregion is successful or not. These are organizational development, diversification of resource base, and appropriation of cross-border activities[21].

Applying these criteria, the EUREGIO has to be judged as being among the more successful cases of CBC in Europe. It has developed a strong, independent base, a diversified stream of income (compared to other Euroregions) and a portfolio of various activities that go beyond the mere administration of Interreg funding.

Have Euroregions generated innovative forms of governance? On the one hand, the EUREGIO must be seen as one of the pioneers driving

[18] A. Cochrane, J. Peck, and A. Tickell, "Manchester Plays Games: Exploring the Local Politics of Globalisation", in *Urban Studies*, Vol. 33, No. 8, 1996, pp. 1319-1336.

[19] M. E. Smyrl, "Does European Community Regional Policy Empower the Regions?", in *Governance*, Vol. 10, No. 3, 1997, pp. 287-309.

[20] M. Perkmann, "Policy Entrepreneurship and Multi-level Governance: a Comparative Study of European Cross-Border Regions", *op. cit.*

[21] M. Perkmann, "Policy Entrepreneurship and Multi-level Governance: a Comparative Study of European Cross-Border Regions", *op. cit.*

the development of the Euroregion model in the first place[22]. At the same time, however, the innovation consisted mainly in developing an organisational form that would occupy the niche carved out by the EU Interreg programme. In other words, the organisational form embodied by the EUREGIO represents an answer to the need of the European Commission for regional policy implementation partners that would be able to devise and initiate cross-border projects on a local level.

Therefore, like many Euroregions, the EUREGIO functions as an implementation agency for this specific type of EU cohesion policy. The European Commission is a policy-maker that has no proprietary implementation apparatus and it therefore relies on external agents. This is the context where organisations such as the EUREGIO can play an important role. By operating at the cross points between its local (horizontal) network of constituencies and the multi-level (vertical) network related to cohesion policy implementation, the EUREGIO's organisational form has been carefully crafted to fit into this overall framework of EU cohesion policy. Both networks have demonstrated to be highly stable and robust over time but, eventually, their durability will be tested in case the Interreg programme will not be renewed in its current form.

On the third point, I discuss the question of the impact of the EUREGIO to date with reference to the discussion as to how CBC can be assessed in more general terms.

The impact of CBC policies on their socio-economic environment is notoriously difficult to assess. As required by EU regulations, the EUREGIO has developed a series of metrics that it uses for assessing the immediate outcome of the Interreg-funded activities. For instance, the evaluation of the EUREGIO Interreg II programme states that during 174 projects involving almost 800 partner organisations more than 1,700 jobs were either directly or indirectly created or 'secured'[23]. Also, more than 75% of Interreg-funded projects are said to be continuing after Interreg funding runs out. These figures have to read against the actual cost of the policy interventions which amounted to approx. EUR 64 m., including all EU, national and local funds. This means a cost of more than EUR 36,000 for each maintained or created workplace was incurred. Whether this can be judged as a success, needs to be assessed on a comparative

---

22 M. Perkmann, "Euroregions. Institutional Entrepreneurship in the European Union", in M. Perkmann, and N. L. Sum (eds.), *Globalization, Regionalization and Cross-Border Regions*, Houndsmills, Palgrave, 2002, pp. 103-124.

23 EU-Gemeinschaftsinitiative INTERREG-II für die EUREGIO Förderung aus dem Europäischen Fonds für Regionalentwicklung. EFRE No. 94/00/10/020 – Endbericht 31/12/2001.

basis with similar structural policy programmes led by the EU and other policy makers and can therefore not be pursued further here.

It is even more difficult to judge the broader impact of CBC within the area concerned. This is because many of the intended impacts of CBC measures will be, on the one hand, intangible and, on the other, effective only over the medium and long term. One possibility is to qualitatively evaluate the achievements of EUREGIO activity over time. Among others, these involved creation of better public transport links across the border, the generation of countless social contacts via project-linked collaboration, efficiency gains via better cross-border access to private and public services and cross-border labour market integration. In view of these achievements, there is no doubt that by promoting and funding these projects, the EUREGIO has done much to accelerate and deepen local cross-border integration.

However, more systematic impact assessment attempts need to be carried on a comparative basis across other CBC cases. To date, there has been little research addressing these broader questions of CBC impact. In addition, the few existing studies remain relatively sceptical. Even where Euroregions have existed for decades, these studies find that efforts to intensify cross-border economic contacts have been successful only to a limited degree[24]. Research on cooperation in the Benelux area and at the German-Polish and Finnish-Swedish borders indicates that the majority of inter-firm linkages are still confined by their national economic spaces[25]. Many observers more generally comment on the barriers that prevent cross-border bodies from being effective and efficient governing units acting on behalf of cross-border units as a whole[26].

---

24 M. v. Geenhuizen, B. v. d. Knaap, and P. Nijkamp, "Transborder European Networking: Shifts in Corporate Strategy?", in *European Planning Studies*, Vol. 4, No. 6, 1996, pp. 671-682.

25 H. v. Houtum, *The Development of Cross-Border Economic Relations*, Tilburg, Center for Economic Research, Tilburg University, 1997; S. Krätke, "Regional Integration or Fragmentation? The German-Polish Border Region in a New Europe", in *Regional Studies*, Vol. 33, No. 7, 1999, pp. 631-641; E. K. Prokkola, "Cross-Border Regionalization, the INTERREG III A Initiative, and Local Cooperation at the Finnish-Swedish border", *op. cit.*

26 J. Beck, *Netzwerke in der transnationalen Regionalpolitik: Rahmenbedingungen, Funktionsweise, Folgen, op. cit.*; A. Church, and P. Reid, "Cross-Border Co-operation, Institutionalization and Political Space across the English Channel", in *Regional Studies*, Vol. 33, No. 7, 1999, pp. 643-655; C. K. Hahn, "The Transboundary Automotive Region of Saar-Lor-Lux: Political Fantasy or Economic Reality?", in *Geoforum*, Vol. 48, 2013, pp. 102-113; E. Nadalutti, "Does the 'European Grouping of Territorial Co-operation' Promote Multi-level Governance within the European Union?", in *JCMS: Journal of Common Market Studies*, 2013, pp. 1-16; J. B. Harguindéguy, and Z. Bray, "Does Cross-Border Cooperation Empower European Regions? The Case of INTERREG III-A France-Spain", in *Environment and Planning C, Government & Policy*, Vol. 27,

Compared to the sceptical tone of some of these assessments, the EUREGIO fares relatively well. The independence the secretariat achieved versus the member authorities means that Interreg-funded projects in the EUREGIO are likely to be more 'cross-border' in nature than in cases where no such cross-border agency exists. The EUREGIO secretariat has developed a self-interest in cross-border matters that – in conjunction with a high level of experience and expertise – guarantee it will more likely represent the interests of the cross-border space as a whole rather than of single participating public authorities. At least procedurally, the EUREGIO can therefore be seen as well equipped for promoting integration within the local cross-border space it occupies.

Concluding, the fact that the EUREGIO, like most CBRs in Europe, is highly dependent on EU cohesion policy support via the Interreg programme is both blessing and curse. On the one hand, Interreg and previous EU support has provided the opportunity for the EUREGIO to consolidate its achievements and establish an important role for itself within the public sector administration in the Dutch-German border area. On the other, the dependence on Interreg puts significant limits on its discretion over priorities and activities. These limits are exemplified by the tensions over the policy priorities relating to Interreg IV, the successor programme for Interreg III from 2008 onwards. While the preference of the EUREGIO and its local members is to run a broad-based portfolio of activities across all 'spheres of life', the Dutch government would like to see a stronger focus on specialised technology and innovation measures in pursuit of the Lisbon Agenda. In this situation, it has to be acknowledged, the existence of the EUREGIO as an established cross-border centre of competence will make a difference by enhancing the bargaining power and influence of the local partners on the general direction of Interreg implementation.

No. 4, 2009, p. 747; E. K. Prokkola, "Cross-Border Regionalization, the INTERREG III A Initiative, and Local Cooperation at the Finnish-Swedish Border", *op. cit.*

Cross-Border (Im)mobility in Times of Crisis

Martin VAN DER VELDE

Radboud University Nijmegen

1. Introduction

Mobility has always been regarded as a major factor in regulating all kinds of regional and global imbalances. Noticeably mobility for economic reasons, be it companies off-shoring their activities or people moving for labour reasons, is supposed to counteract inequalities, both where it concerns the supply and the demand side. Especially in times of crisis, which most of the time exposes itself with several levels of intensities in different regions, mobility is supposed to even out the effects. Focusing on the European Union, it is clear that great differences do exist when looking at the effects of the recent financial crisis. Unemployment rates, particularly amongst the youth are far higher in Southern European countries, compared to many of the countries in Northern Europe. One could suppose that in the European Union, where one of its pillars is the free movement of people, this should result in strongly increasing levels of labour mobility.

Mobility within the European Union has always been encouraged as a major goal to the economic integration process[1]. If we look at the current situation, it is not bold to state that this cohesion has been challenged in the last decade. But even before this era of crisis, mobility was regarded as a major instrument. This has been so much so that stimulating mobility has become almost a goal in itself. In this respect, mobility in the United States has often been considered to be exemplary. EU citizens have about half the mobility rate of US citizens. A little under 40% has moved house in the last 10 years. The great majority of them moved within the same town or village. Only one in 25 moved to another member state[2]. Also, if

---

1 G. Gottholmseder, and E. Theurl, "Determinants of Cross-Border Commuting: Do Cross-Border Commuters within the Household Matter?", in *Journal of Borderland Studies*, Vol. 22, 2007, pp. 97-112.

2 EC (European Commission), *High Level Task Force on Skills and Mobility: Final Report*, Brussels, European Commission – Directorate-General for Employment and Social Affairs, 2001, http://bookshop.europa.eu/en/high-level-task-force-on-skills-

we compare mobility in the EU (as a seemingly 'borderless' world) with the levels witnessed in China, it is extremely low. In 2007 almost half of the urban workforce consisted of migrant workers[3].

It is probably not completely fair to compare the EU-levels with the inter-state mobility in the US and the internal mobility in China, as part of this mobility (certainly in China) is not voluntary. But nevertheless, the EU considers its mobility levels to be too low. 2006 was even declared as the "Year of Workers Mobility". This nicely fitted the motto stemming from one of the Communications of the Commission stating that European labour markets should be 'open to all, with access for all', in order to 'reduce pressures to migrate for those who do not want to move, while creating genuine opportunities for those who do wish to be mobile'[4].

In the last decades of the previous century and the early days of this century much effort has been devoted to stimulate mobility, without too much apparent success. From a recent Eurobarometer comes that only 17 percent of the EU citizens envisage working in another country in the future[5]. As mobility levels in general stayed still relatively low, member states basically supported these EU-policies rather unproblematically. Where the policies of the European Union have had only moderate success, it seems that the current financial and economic crisis has had much more effect. Especially on the countries where the crisis has hit the hardest, outward mobility increased (at the same time still receiving considerable numbers of immigrants from Central and Eastern Europe). So in general mobility levels seem to be on the rise. Currently member states and associated countries are showing tendencies to become more reluctant to embrace mobility goals fully. Switzerland, for instance, announced in late April 2013 that it would use its "pressure valve clause" (*Ventilklausel*), to regulate the influx of people from the EU.

This chapter aims to evaluate the development of EU cross-border mobility and immobility in the current era, which is plagued by crises. The focus will be on the EU. Before presenting a quantitative analysis, it will elaborate on the causes and motives, especially on an individual level. We will discuss an approach that has been put forward recently, in

---

and-mobility-pbKE4302082/ [accessed on 25th November 2013]; EC (European Commission), *Geographic Mobility in the European Union: Optimising its Economic and Social Benefits*, Brussels, European Commission, 2008.

3 C. Fang, D. Yang, and W. Meiyan, *Migration and Labor Mobility in China*, New York, UNDP, 2009.

4 EC (European Commission), High Level Task Force on Skills and Mobility: Final Report, *op. cit.*, p. 5.

5 EC (European Commission), *Special Eurobarometer 337 – Geographical and Labour Market Mobility*, Brussels, European Commission, 2010.

which we have tried to put mobility beyond a neo-liberal rational choice perspective[6].

2. An Individual Threshold-approach

To be able to understand (im)mobility patterns, it is our conviction that it is important to take the stance of the (im)mobile decision making individual. Mobility-patterns, after all, consist of individual actors. Secondly, it is important to note that the majority of people still stay immobile, even if they are living in unfavourable circumstances. Taking into account these premises, we propose an approach to mobility in which (non)decision making in regard to several thresholds is central. Considering these thresholds is also an expression of the fact that we do consider mobility not as a simple Euclidian 'as the crow flies' move from a place of origin to a place of destination. Next to these origin-destination aspects, also an individual state of mind or disposition towards mobility has to be taken into account; and thirdly, the fact that mobility also involves a trajectory[7]. What this approach is trying to understand is why people finally end up where they are if they at all are becoming mobile. In the next sections we will elaborate on the two pillars in this approach, decision-making and the thresholds themselves.

2.1. Decision Making

Mobility in general is very often still seen as an outcome of a rational decision of an individual. Building on assumptions in the seminal work by Ravenstein on migration[8], these structural approaches consider the actual and potential mobile people as atomised particles exposed to and moving in a field of demand and supply forces. Ravenstein, in his original approach, only took wage differences as underlying principle, were distance was the controlling variable. It was basically a distance decay model, were the differences in wages as a pulling force are offset by the distance someone has to cover to take advantages of these differences. As long as the outcome of the equation is positive, the latent mobility of an individual should become manifest. Later on, this simple model has been modified and adapted, first of all taking into account the fact that

6 M. van der Velde, and T. van Naerssen, "People, Borders, Trajectories: an Approach to Cross-Border Mobility and Immobility in and to the European Union", in *Area*, Vol. 43, No. 2, 2011, pp. 218-224.

7 M. van der Velde, and T. van Naerssen, "People, Borders, Trajectories: an Approach to Cross-Border Mobility and Immobility in and to the European Union", *op. cit.*

8 E. Ravenstein, "The Laws of Migration", in *Journal of the Statistical Society of London*, Vol. 48, 1885, pp. 167-235; E. Ravenstein, "The Laws of Migration", in *Journal of the Statistical Society of London*, Vol. 52, 1889, pp. 241-305.

mankind is not omniscient, but is better characterised as using a bounded rationality[9]. Another factor that was put into the equation was the issue of human capital[10]. This factor links the decision about becoming mobile or not to learned and learning experiences. Also, the notion of transaction costs[11] is generally accepted to play a role, certainly when the decision to become mobile is seen as some kind of an economic decision. And last but not least, in the 'New Economics of Labour Migration'[12], the focus in migration shifted from the individual per se to the individual as part of a household, family or even community.

But even if we take all these sophistications into account, the approaches still use a basic push-pull model, trying to explain why people should or would become mobile. This does not really appreciate the fact that the large majority of the population, even in times of crisis, is immobile. Witnessing this, Straubhaar and his colleagues in the late 1980s of the previous century developed the so-called *insiders advantage approach*[13]. The innovation in this approach was that also a value was attached to staying immobile. This can be labelled as a keep-factor. For instance, the loss of specific competences build up in a specific job at a specific place might prevent someone from moving to another employer. In general, this effect is even bigger when moving to a job in another country. Straubhaar focussed very much on this kind of immobility factors at the original place of work, but likewise there can be repel-factors, connected to the potential new place of work that may prevent a move[14]. Of course, as with the push and pull factors, these keep and repel factors are closely connected to each other. And what is more, it still fits into the tradition of rational decision making, where a potential mobile individual is weighing the pros and cons of moving, and based on the outcome decides to move or not.

In the threshold approach, we can argue that it is not always the case that people consciously or unconsciously are willing to make these

9 H. Simon, *Approaches of Bounded Rationality: Behavioral Economics and Business Organization*, Cambridge MA, MIT Press, 1982.

10 G. Becker, "Investment in Human Capital: a Theoretical Analysis", in *The Journal of Political Economy*, Vol. 70, 1962, pp. 9-49.

11 O. Williamson, "The Economics of Organization: The Transaction Cost Approach", in *The American Journal of Sociology*, Vol. 87, No. 3, 1981, pp. 548-577.

12 O. Stark, "Research on Rural-to-Urban Migration in Less Developed Countries: the Confusion Frontier and why we should Pause to Rethink Afresh", in *World Development*, Vol. 10, 1982, pp. 70-3.

13 T. Straubhaar, On the Economics of International Labour Migration, Bern, Haupt, 1988.

14 H. van Houtum, and M. van der Velde, "The Power of Cross-Border Labour Market Immobility", in *Tijdschrift voor Economische en Sociale Geografie*, Vol. 95, 2004, pp. 100-107.

'rational' decisions in regard to mobility. This is attributed to the power of the feelings of belonging. In a spatial sense, these feelings can be projected on a specific bordered space, which then becomes the main frame of reference. Very often the national state acts as this reference, leading to a certain level of (manifest or latent) indifference towards the other side of the border[15]. When this happens, it can lead to an inactive or passive attitude with regard to including these spaces in the 'rational' decision-making process[16]. In a sense this indifference causes some kind of mental threshold that has to be overcome before someone is entering an 'active rational' phase of decision making. Here of course, new thresholds can come into existence. And this mental indifference does play out in all the dimensions of mobility that are discerned, the individual disposition, the locational aspects of the origin and destination and the aspects involving the trajectory. In the next section, we will elaborate on these three issues.

2.2. Thresholds

In our approach to mobility, as mentioned, we discern explicitly mental thresholds with regard to three dimensions, the individual disposition, the destination and origins for mobility and the route that has to be taken to move from the origin to the destination, in other words, people, locations and trajectories. In all of these aspects borders do play a role, albeit in different ways.

The mental threshold embedded in the individual disposition towards mobility is caused by the fact that mankind, to a large extent, is inclined to stay immobile. As mentioned before, a very small minority of the world population has become mobile in the sense that they are currently living in a country different from the one where they were born. It has been estimated that at the world level, only 3 per cent of the population lives outside the country where they were born[17]. Of course this cross-border mobility is only a part of the total mobility, also including moving within one's country. But also on these lower levels of the scale, it can be shown that longer distance mobility is not the norm. In 2005 one third of the mobile people moved house within the same town or city, a quarter within the same region and less than 20% to another region, but within the same

15 M. van der Velde, M. Janssen, and H. van Houtum, "Job Mobility in the Dutch-German Regional Labour Market. The Threshold of Indifference", in G. van Vilsteren, and E. Wever (eds.), *Borders and Economic Behaviour in Europe*, Assen, Van Gorcum, 2005, pp. 77-96.

16 H. van Houtum, and M. van der Velde, "The Power of Cross-Border Labour Market Immobility", *op. cit.*

17 United Nations, Department of Economic and Social Affairs, Population Division, *The Age and Sex of Migrants 2011 Wall chart*, 2011 (United Nations publication, Sales No. 12, XIII.2).

country[18]. This can be explained by the fact that basically humans are trying to avoid uncertainty and risks. And as moving to another, often less well known, place involves exactly this uncertainty, mobility in general is mostly not the first option if it is considered at all. Although we do not want to present the thresholds in a univocal time-sequential order, it seems logical to suppose that, in order to become mobile, first there has to be a disposition that mobility is an option at all. Having said this, it might be that, even though in an earlier stage this threshold has been overcome, in a later stage an unfavourable disposition might become prevalent again.

The second mental threshold is connected to the location that could be involved in a move that is considered. Before entering a location in a more or less rational decision making process of weighing pros and cons, it has to be considered and recognised as an option at all. In other words it has been on the 'long list'. A mental threshold with regard to certain locations may prevent this. As we realise that people are not really inclined to become mobile, certainly not over longer distances, these mental thresholds might be higher for locations that are further away and much more unfamiliar. And, already preluding the role border may play in this process, cross-border locations are very often more unfamiliar and mentally also more distant. Again, to reiterate that we do not want to present this approach as time-sequential, we could suppose that a locational trigger is preceding a stimulus for moving at all. So it is not the wish to move, but the possibilities or the allure of a certain location that triggers the decision-process.

The third threshold we do recognise in our approach is connected to the trajectory that has to be covered in order to move from the origin to the destination. This factor is more prevalent and discussed in the context of long distance international and undocumented migration, where the route of passage very much determines what the (temporary) destination of these migrants will be. But also looking at long distance migration flows of documented migrants awareness of routes and trajectories is quite determining. Here we can state that unawareness of certain routes, and there with possible destinations, can also be labelled as some kind of indifference. So here it is not so much the unawareness of the destination that is the prime reason for not being included in the decision making process, but much more the fact that a possible trajectory is outside the window of considered possibilities. For instance, if one knows that a particular Greek island is a wonderful location, but is unaware of the possibilities to get there, this location will not be included in the process.

18 Eurofound (European Foundation for the Improvement of Living and Working Conditions), *Mobility in Europe*, Luxembourg, Office for Official Publications of the European Communities, 2006.

As stated, trajectories might at first be considered more important in long distance migration, but also when studying shorter-distance migration and commuting, trajectories can play a role. Especially in a cross-border context, the possibilities for travelling between places are not always known. This can also hamper the changes of cross-border opportunities for work to be recognised. And vice-versa, 'well-serviced' routes might favour certain locations at the expense of others. For example in the early days after the accession of Poland, many came to the Netherlands in a circular migration pattern, as this was facilitated very well by employment agencies, especially in the horticultural sector[19].

2.3. Borders

In the previous section, we already mentioned some aspects of borders that are of importance when discussing the relevance of the several thresholds. In regard to the personal attitude, where the feelings of belonging and avoiding uncertainty and risk are relevant factors, the national state plays a major role. State and national institutions, like education and media, but also national systems like health, welfare and social security are all very much favouring the national territorial delineation. The frames of reference for belonging, the knowledge framework, and zones of comfort, very much coincide with the national territory. The national border is then the demarcation of these zones. What is beyond is in many cases a zone of indifference (consciously or unconsciously). Mobility across this border of comfort then is not considered.

With regard to the locational threshold, this is also creating a set of opportunities that are not considered when the first threshold has been taken of moving at all. Next to not being on the long-lists of the decision maker, if a cross-border location is considered, also the evaluation of the pull and repel factors might get a different weight in the deliberation process.

Finally, as already mentioned, borders may play a role when it concerns trajectories. On the one hand, as with the locational aspect, the mere knowledge of the existence of cross-border opportunities may be missing, but also using a cross-border connection might be considered more troublesome compared to national routes. And of course borders as national institutions might be considered as a physical barrier on the projected trajectory and so making it impossible to become mobile across the border.

---

19 R. Pijpers, and M. van der Velde, "Mobility across Borders: Contextualizing Local Strategies to Circumvent Visa and Work Permit Requirements", in *International Journal of Urban and Regional Research*, Vol. 31, 2007, pp. 819-835.

3. EU Cross-Border Mobility in the Early 2010s

In this section, we will try to paint the picture of cross-border mobility in the EU. The fact that we want to get a comparable comprehensive image for all EU countries immediately confronts us with limitations because of the availability of data. When defining cross-border mobility two major forms can be discerned, to wit migration and commuting. In this analysis we will confine ourselves to cross-border migration. One of the reasons is that from earlier studies[20] comes that cross-border commuting is (still) a relatively small phenomenon. Defining a border-region at the Eurostat NUTS-2 level, it came that in 2001 in the EU-15 about 500,000 workers commuted across a national border out of a total workforce of 34 million that is about 1.5 per cent. At that time, the countries of the EU-15 accommodated about 3 million foreign-born workers from another EU-15 country. Of course with the enlargement of the EU, the absolute number of cross-border commuters will have increased. In 2008 in the EU-15, 1.7 per cent of the workers in border regions commuted across the border. If the 12 new member states are added, this share is 2.1 per cent[21]. However, commuting is considered to be too low and rather exceptional. It remains newspaper worthy when specific initiatives are undertaken by employment agencies in border regions to get people interested in jobs on the other side of the border. A more practical reason is that if at all available data on cross-border commuting on an EU-level is hardly comparable and not very reliable.

Also, when we confine ourselves to migration, there are two major groups to be discerned, permanent and temporary migration. In regard to the latter, several studies have concluded that although temporary migration is substantial, data on the size and duration of these forms of migration are unreliable and also inconclusive[22]. Although recognizing the importance of this phenomenon, in our attempt to create an EU-wide picture, again because of availability of reliable data we will limit our analysis to the first category of more permanent migration. The advantage of this group is that they have to register themselves and are included in the national statistics more comprehensively.

For our purposes two sources will be used. As a first indicator the share of inhabitants with a foreign citizenship is incorporated. Using this indicator,

---

20 MKW, *Scientific Report on the Mobility of Crossborder Workers within the EEA*, Munich, MKW, 2001; H. van Houtum, and M. van der Velde, "The Power of Cross-border Labour Market Immobility", *op. cit.*

21 MKW, Scientific Report on the Mobility of Cross-Border Workers within the EU-27/ EEA/EFTA Countries, Munich, MKW, 2009.

22 See B. Galgóczi, J. Leschke, and A. Watt (eds.), *EU Labour Migration in Troubled Times: Skills Mismatch, Return and Policy Responses*, Farnham, Burlington, VT, Ashgate, 2012.

we have to keep in mind that these figures are underestimating the shares somewhat as they do not take into account illegal flows. But we do not expect this to be significant when looking at inner-EU mobility. As mobility is put mainly in the context of the labour market, the age-group of 15-64 will be used here. This indicator is first and foremost measuring the (rather static) stock of non-nationals. To also get a more dynamic picture, the international migration flows will be used as a second indicator. A disadvantage of this statistic is that it is far less comprehensively available for the countries of the EU. This data therefore will perforce be used to illustrate the dynamics in the recent period for a selection of the EU-countries.

3.1. Foreign Citizenship

In 2011, the EU-28 had a total population in the age of 15 to 64 of a little over 340 million. Of these, a total of about 26 million (7.8 per cent) did possess a citizenship other than the country of living, and 16 million from a country other than the EU-28 (table 1). About 3 per cent of the population between 15 and 64 can be said to have been mobile within the area of the EU-28. This share is slowly growing. Since 2009 (the earliest year with comparable data available), it has increased by about 0.2 per cent. If we take the comparable statistic of the share of the EU-28 citizenship in the total population, we can see that it has increased from 1.9 per cent in 2003 to 2.5 per cent in 2011 and 2.7 in 2012.

Table 1. Citizenship of the Cohort 15-64 in the EU-28 in 2011

			EU28		**Other**	
	Tot.	**National**	**Tot.**	**%**	**Tot.**	**%**
Austria	5,682,292	4,969,200	275,400	*4.8%*	437,692	*7.7%*
Belgium	7,250,159	6,367,314	556,803	*7.7%*	322,617	4.4%
Bulgaria	5,033,849	5,001,270	6,622	0.1%	25,898	0.5%
Croatia *	4,436,401	4,399,301	7,800	0.2%	29,300	0.7%
Cyprus	591,923	454,039	80,046	*13.5%*	57,838	*9.8%*
Czech Republic	7,378,802	7,012,401	120,681	1.6%	245,720	*3.3%*
Denmark	3,631,760	3,353,125	106,061	2.9%	172,452	4.7%
Estonia	896,436	743,017	10,469	1.2%	142,950	*15.9%*
Finland	3,546,558	3,411,113	50,006	1.4%	84,769	2.4%
France	42,197,325	39,493,933	924,544	2.2%	1,778,848	4.2%
Germany	53,966,108	48,195,478	2,153,355	4.0%	3,617,275	*6.7%*
Greece	7,505,316	6,730,070	143,198	1.9%	632,048	*8.4%*
Hungary	6,857,377	6,684,088	105,104	1.5%	68,185	1.0%
Ireland	2,983,602	2,638,525	250,331	*8.4%*	55,282	1.9%
Italy	39,811,683	36,210,798	1,117,484	2.8%	2,483,401	*6.2%*

			EU28		Other	
	Tot.	**National**	**Tot.**	**%**	**Tot.**	**%**
Latvia	1,535,986	1,273,493	8,073	0.5%	254,420	*16.6%*
Lithuania	2,220,937	2,194,614	1,551	0.1%	24,772	1.1%
Luxembourg **	342,909	183,267	137,366	*40.1%*	22,276	*6.5%*
Malta	289,070	274,478	6,772	2.3%	7,820	2.7%
Netherlands	11,153,778	10,529,093	280,520	2.5%	279,253	2.5%
Poland	27,256,968	27,214,436	11,826	0.0%	26,952	0.1%
Portugal	7,097,788	6,714,794	83,505	1.2%	299,489	4.2%
Romania ***	15,036,386	15,008,065	5,001	0.0%	23,150	0.2%
Slovakia	3,932,092	3,871,627	37,684	1.0%	22,781	0.6%
Slovenia	1,420,392	1,347,849	4,308	0.3%	68,235	*4.8%*
Spain	31,310,040	26,817,768	1,815,339	*5.8%*	2,676,933	*8.5%*
Sweden	6,113,365	5,626,620	204,391	*3.3%*	274,726	4.5%
United Kingdom	41,177,537	37,609,711	1,566,716	*3.8%*	2,001,110	4.9%
Total	*336,220,438*	*309,930,186*	*10,063,156*	*3.0%*	*16,106,892*	*4.8%*

* Croatia: 2008 and total population.

** Luxemburg: 2010.

*** Romania: 2009.

Italic: above EU average.

Source: Eurostat 2013 and own calculations.

If we break this total down for the different countries, there are quite marked differences to be seen. The share of foreign EU-citizens ranges from almost nothing in some of the newly acceded countries as Poland, Romania, Bulgaria and Lithuania, to more than 5 per cent in Spain, Belgium, Ireland, Cyprus and Luxemburg. The high share of Belgium is caused by the importance of the international (i.e. European) institutions. Luxemburg traditionally is an important host country for Portuguese. 16 per cent of the total population (or 40 per cent of the foreign population) of Luxemburg is of Portuguese descent[23]. The shares in Ireland and Cyprus can partly be explained from the historical roots with the UK. For instance, 28 per cent of the EU-28 foreign citizens in Ireland are British. Next to that, Ireland also has seen a large increase in Poles, because they did not impose any transition measures after the 2004 accession round. For Cyprus a national breakdown is not available but next to Greek citizens, (historically) it also accommodates a large group of Brits. Spain has seen a major group of Romanian citizens moving to its territory.

[23] http://www.statistiques.public.lu/fr/publications/series/rp2011/2013/18-13-portugais/index.html [accessed on 26th November 2013].

As a small sidestep, we do also see some interesting patterns when looking at the non-EU foreign citizens. The highest and the lowest shares are here located in the newly acceded countries in Central Europe. Estonia and Latvia have the highest shares, mainly Russians, also for historic reasons. With the exception of the Czech Republic and Slovakia the other Central-European countries do show low shares of non-EU citizens.

Based on the Eurostat-data it is possible to unravel the aggregate data for some of the individual countries and distil the most important countries when it comes to foreign EU-28 citizenships (table 2).

Table 2. Most Important Foreign Citizenships per Country (> 10% of all Foreign EU Citizens)

Hosting country	Citizenship
Belgium	Italy (22%); **France** (19%); **the Netherlands** (18%)
Czech Republic	**Slovakia** (52%); **Poland** (13%); **Germany** (10%)
Denmark	Poland (18%); Germany (17%); Slovenia (10%); United Kingdom (12%)
Germany	Italy (19%); **Poland** (16%); Greece (10%)
Ireland	Poland (29%); **United Kingdom** (27%); Lithuania (12%)
Spain	Romania (36%); United Kingdom (16%)
Italy	Romania (71%)
Latvia	**Lithuania** (38%); Germany (12%)
Hungary	**Romania** (61%); Germany (16%)
Netherlands	**Germany** (21%); Poland (16%); **United Kingdom** (12%)
Poland	**Germany** (30%)
Portugal	Romania (36%); United Kingdom (17%)
Slovenia	**Croatia** (59%)
Slovakia	**Czech Republic** (21%); Romania (14%); **Hungary** (13%); **Poland** (13%)
Finland	Slovenia (14%)
Sweden	**Finland** (26%); **Denmark** (15%); Poland (15%); Germany (10%)
United Kingdom	Poland (35%); **Ireland** (22%)

Bold: neighbouring country.

Hosting countries that are not listed do not report on the breakdown of foreign citizenships.

Source: Eurostat 2013 and own calculations.

First of all, it can be seen that in most of the countries that report on citizenship, a few countries make up the bulk of the inhabitants with a foreign citizenship. Furthermore, neighbouring countries are important when it comes to this kind of mobility. These are already first indications that mobility in the EU can be characterised as very concentrated and relatively short distanced. In terms of thresholds, the locational threshold seems here at work, first of all, in favouring certain historical familiar 'distant' countries at the expense of others. Secondly, other more distant countries are generally less well known and therefore less often considered as possible destinations compared to neighbouring countries.

Finally, when looking at the growth rates since 2009 we do see quite large differences in the EU-28 (table 3). Lithuania is losing both EU-28 citizens as well as citizens from outside the EU. We do have to keep in mind here that for many countries the absolute levels were quite low, so that small absolute changes can express themselves in considerable percentages.

Table 3. Change in the Absolute Numbers of Foreign Citizens between 15-64 years old in the EU-28 between 2009-2011

	EU-28	Other
Austria	9.1%	2.2%
Belgium	7.4%	30.2%
Bulgaria	113.3%	44.4%
Croatia	:	:
Cyprus	29.2%	33.0%
Germany	3.2%	-0.9%
Greece	-1.2%	2.8%
Hungary	14.8%	7.9%
Ireland	0.9%	1.3%
Italy	20.5%	16.3%
Latvia	4.8%	-6.7%
Lithuania	-25.6%	-18.6%
Luxembourg*	-0.2%	16.5%
Malta	23.6%	-2.4%
Netherlands	14.7%	-0.8%
Czech Republic	-9.0%	6.0%
Denmark	14.2%	6.0%
Estonia	34.5%	-4.2%
Finland	17.9%	17.6%
France	1.1%	1.1%
Poland	-17.9%	8.7%
Portugal	21.4%	-3.0%
Romania	:	:
Slovakia	27.2%	30.8%
Slovenia	30.4%	14.8%
Spain	1.2%	-3.6%
Sweden	4.1%	21.4%
United Kingdom	12.8%	1.9%
EU-28	6.8%	2.9%

* Luxemburg: 2010-2011.

Source: Eurostat 2013 and own calculations.

Poland is losing to the EU-28 but still seems to be attractive for non-EU citizens. The same goes for the Czech Republic, albeit on a more moderate level. Greece is also losing to the EU, which might be crisis related, but is still witnessing slightly increasing numbers from outside the EU. Luxemburg does show an interesting pattern, where the number of EU citizens is relatively stable, but where non-EU citizens do show a considerable increase. On the other side of the spectrum, most of the other Central European countries do see the number of EU-citizens grow. Again we have to keep in mind that absolute numbers here are still not too impressive. Although not at the core of this analysis, it is interesting to see that some countries are getting less inflow from outside the EU. Worth mentioning here is especially Spain and to a lesser extent Portugal. Both of them have seen the number of non-EU citizens diminish. This can be partly explained by a relative economic boom in South-America, in particular Brazil.

3.2. Migration Flows

The previous section tried to take inventory of the stock of (im)mobile people as far as it can be deduced from 'foreigners' living in other EU-countries. To add to a more dynamic aspect that was already introduced by the growth rates of foreign citizenship in recent years, in this section, the migration flows will be used to further analyse the recent mobility patterns in the EU. As it is not possible to analyse all 28 countries, a selection was made. This selection is partly guided by the availability of consecutive data series for the early years of this millennium. Together with the aim to trace the effects of the crisis and other transitions on mobility patterns in differently affected countries, this led to include Denmark, as an 'average' European country, Lithuania, as an accession country and Spain, as a country severely affected by the Euro and financial crisis.

Before analysing these three countries in more depth, some general observations can be made. The first one concerns the fact that very often there is a quite strong reciprocity in the migration flows of countries. That is, countries that are foci of immigration are most of the time also important as the destination for emigration or are becoming so after a certain time lag. This could point in the direction of the importance of return migration. For instance in Germany between 2004 and 2008 immigration from Romania rose from around 24,000 to about 48,000. At the same time, emigration increased from almost 20,000 to 38,000. In the UK between 2008 and 2011 on average 40,000 arrivals from Poland were registered, but there were also 30,000 departures towards Poland. From the statistics on citizenship comes that 33,000 Polish citizens left the UK. Although it is not possible to link citizenship and next usual residence after migration directly, it seems reasonable to suppose that a large share of Polish citizens migrate back to Poland as their next place of living, which would fit the idea that some form of circular migration is gaining importance in the EU.

A second general observation is that mobility in general has increased. Of course in, and shortly after, the accession years 2004 and 2007, the increase is partly caused by migration from central European countries. The fact that mobility is staying at a relatively high level thereafter might also be evidence of the earlier mentioned return migration. When we look at the composition of the emigration flows from the EU-15 countries (as far as they are available) we can see that in all of these countries, except Germany and Luxemburg, the share of nationals in the emigration flows have decreased between 2002 and 2011.

Based on Eurostat-data, it is possible to disaggregate the migration-patterns at a quite detailed level. In the next sections, Denmark, Lithuania and Spain are used as exemplary cases. For these countries the 'bilateral' relations between 2002 and 2011 are analysed for those countries that had a share of more than 10% in either the immigration or emigration flow in at least one year.

3.3. Denmark

Economically, Denmark has performed reasonably well during the Euro- and financial crisis. Of course it is also affected and has experienced declining economic growth rates and increasing unemployment rates but in comparison to many other EU-countries this has been relatively modest.

Figure 1. *Total Migration Flows in Denmark.*

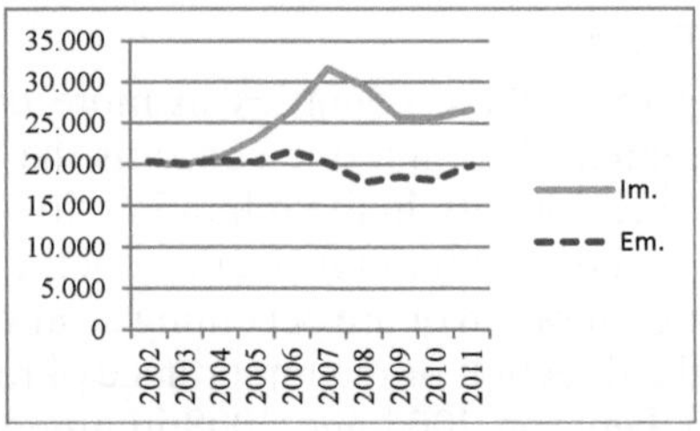

The total migration flows have remained relatively stable during this period (Figure 1). Emigration is hovering around 20,000 per year. Immigration is a bit more volatile, being around 20,000 in the early years, increasing to about 30,000 around 2008 and recently levelling around 25,000.

The most important countries when it comes to migration are Poland, Sweden, Germany, the UK and recently Romania. As said before, migration flows do exhibit a quite remarkable reciprocity. Looking at the development over time, for all five countries but Germany the patterns are quite congruent, albeit with a certain time lag (Figure 2). Until 2008 Denmark did witness increasing immigration from all countries, except the UK. Denmark was especially interesting for Polish migrants. Since 2008 immigration diminished for all countries with the exception of Romania, which of course is understandable because of the more recent accession to the EU. What is remarkable here is the continued rate of the increase. Romanian migration towards many of the other EU-countries has slowed down in recent years or even declined, as we will see in the case of Spain. The spike in immigration from Germany around 2007 coincides with the slow economic performance of that country in the decade and a half after the reunification.

Emigration, as mentioned, is largely mirroring immigration, mostly at a somewhat lower level. For Poland, for instance, emigration peaked in 2009 at around 3,000 to fall back to approximate 2,000 in recent years. Also when we look at emigration to Romania, we do see some mimicking of the deviant trend here, with a continuing increase until 2011. As we have also mentioned earlier, this emigration seems to be to a large extent return migration. From the Eurostat-data comes that 720 people emigrated from Denmark to Romania, and the 701 emigrants leaving Denmark were Romanian citizens. Of course, it is unknown whether all these Romanians moved to their home country, but it seems reasonable to assume that a large majority did do this. Again as mentioned earlier, this supports the idea of increasing circular trends in migration. The same can be said for emigration to Poland. The most recent available data shows that in total 1992 people migrated from Denmark to Poland, whereas 1929 of all emigrants were Polish citizens.

Figure 2. *Binational Migration Flows from and to Denmark.*

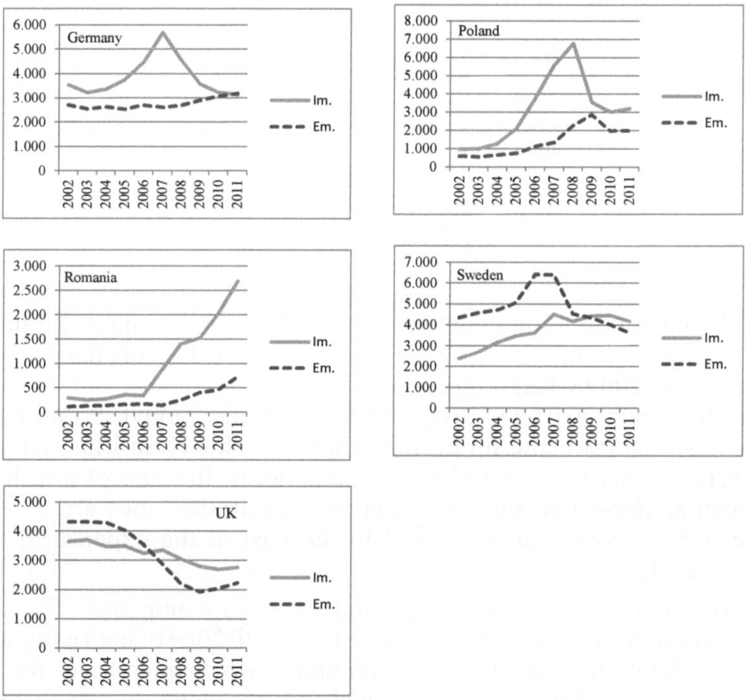

The developments of the flows between Denmark and Sweden and the UK, are in accordance with the generally observed trend that shows a 'reversal' around 2007-2008. In Sweden the growth turned into a stabilisation (immigration) or decline (emigration). In the UK emigration started to decline around this time.

3.4. Lithuania

Lithuania as a country has been doing quite well during the first years of this millennium. It was hard hit by the financial crisis around 2009. This clearly shows in the overall migration figures (Figure 3). Until 2009, emigration moved around a modest 10,000 migrants per year. Unlike many of the other accession countries, emigration did not increase that much after 2004. However this changed dramatically after 2009, with 70,000 emigrants in 2010. In 2011 the number of emigrants declined again, which coincides with the remarkable recovery of the economy in that year. Immigration has been relatively small over this period. Only in 2011 an increase can be seen. Again it is highly probable that this can be labelled as return migration.

Figure 3. *Total Migration Flows in Lithuania.*

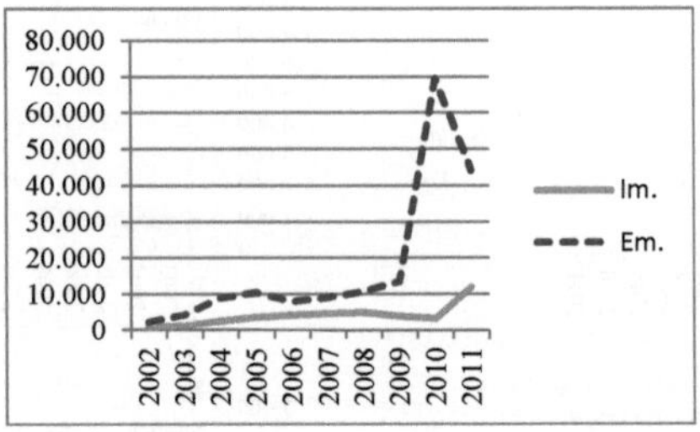

The most important countries for migration from and to Lithuania are the UK, Ireland, Germany and Spain (Figure 4). This has been the case before 2009 when migration was relatively modest, and remained so in the years 2010 and 2011, when migration surged. So no major new destinations came up during this time. This is an illustration of the relative inertia in spatial migration patterns. In terms of thresholds we can suppose that the individual and locational ones are at work here, which favour options used in the past at the expense of new opportunities.

For all the major destinations and origins for migration the same patterns can be seen. A relatively stable level until 2009 of both emigration and immigration, with outward flows about two to three times the size of the incoming flows. Due to the crisis hitting, emigration exploded in

2010. In 2011, for all destinations but Germany, the number of emigrants declined again significantly.

Immigration rises in 2011 from all destinations. As mentioned before, this can be labelled as return migration. Lithuanians in this respect seem to exhibit a somewhat different behaviour though. For Ireland, for instance, the data shows that in 2011 about 2,000 migrants moved to Lithuania. If we look at the number of Lithuanian citizens emigrating from Ireland, this is much higher at about 3,000. Apparently a part of the Lithuanians move on to other countries. This might be caused by the fact that Ireland itself is also hard hit by the crisis, so many migrants are forced to look for opportunities elsewhere. In the case of Lithuania many are not ready to go back to their home country.

Figure 4. *Binational Migration Flows from and to Lithuania.*

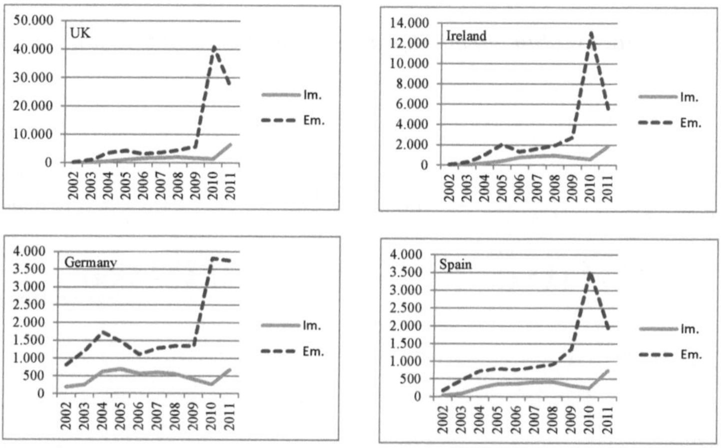

3.5. Spain

Spain is representing the countries in Southern Europe that have been hit by the financial crises the longest and hardest. This is clearly visible in the overall levels of migration (Figure 5). Spain having been an immigration country for a long time, now shows immigration and emigration at about the same level. The inflow to Spain is back at the level it was around 2002. In this year, Romanians (as the most important immigration country) got freedom of travel within the Schengen area. Immigration peaked at 350,000 in 2007 to drop back to 150,000 in 2009. Since then it has remained relatively stable. Emigration is increasing dramatically since 2009. It rose from around 20,000 to 170,000 in 2011.

Figure 5. *Total Migration Flows in Spain.*

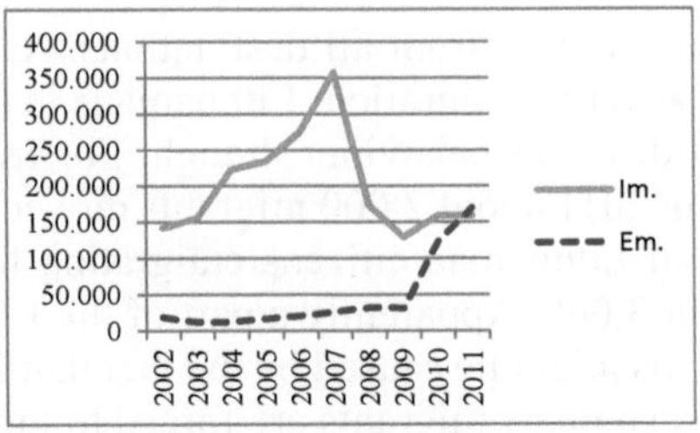

The most important migration countries apart from Romania are Germany, France and Bulgaria. When we break down the overall picture for these countries, some interesting observations can be made (Figure 6).

Figure 6. *Binational Migration Flows from and to Spain.*

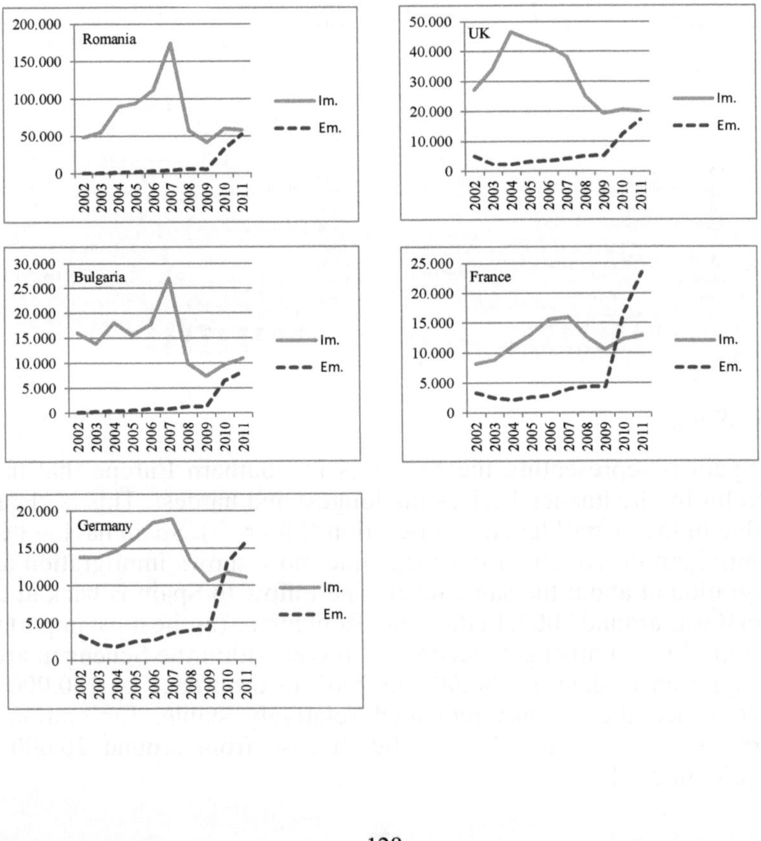

First of all it becomes clear that immigration from all the major migration countries have declined. From the UK this started already in 2004, so well before the crisis. The other countries started to send fewer migrants from 2007 onwards. Emigration rose considerably from 2009. Although again it cannot be deduced univocally from the statistical data available from Eurostat, it can reasonably be supposed that it concerns two different groups. The rising emigration towards Romania and Bulgaria can be characterised more and more as return migration. In the earlier years, just as was observed in Ireland with the Lithuanians, especially the Bulgarians, when emigrating from Spain seemed to be moving to another country rather than their home country. In the last two years, however, the numbers of emigrants to Romania and Bulgaria and the Romanian and Bulgarian citizens that emigrate from Spain are almost equal (table 4).

Table 4. Emigration from Spain

	2011	2010	2009	2008	2007	2006	2005	2004	2003	2002
Total emigration from Spain to Bulgaria	9.063	6.358	1.321	1.276	840	843	545	420	216	121
Bulgarian citizens' emigr. from Spain to any dest.	9.058	6.037	4.975	2.558	1.544	4.030	1.027	977	444	103
Total emigration from Spain to Romania	51.279	32.680	25.901	14.000	8.050	16.446	4.044	3.148	1.438	234
Romanian citizens' emigr. from Spain to any dest.	52.480	33.817	5.673	5.854	4.078	3.433	2.093	1.316	656	271

Source: Eurostat 2013.

A second group concerns the Spanish nationals that emigrate. Whereas in most countries the number of emigrating nationals is rather stable, Spain has witnessed a steep increase since 2004 (Figure 7). As we suppose that many of the immigrants from the central European countries return to their home country, we may conclude that the increase of emigration from Spain to France and Germany (and to a lesser extent the UK) will largely consist of Spanish nationals, looking for jobs in these countries.

Figure 7. *Nationality of Emigrants from Spain.*

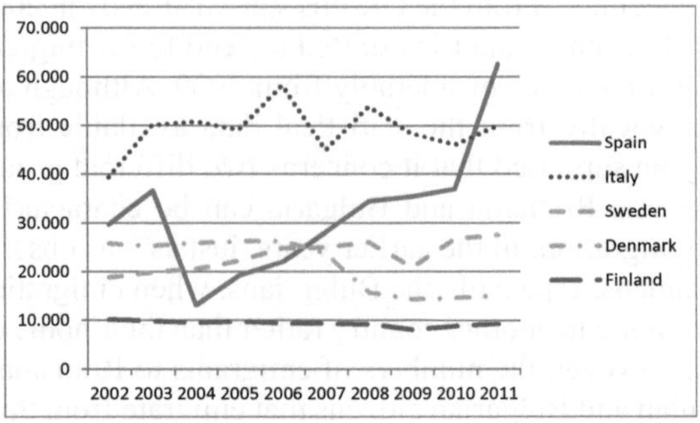

3.6. Summarising

Through these three country vignettes, it seems we can witness three different patterns. Denmark, posing for countries that are experiencing the crisis, but in a moderate sense, mobility, both inward and outward is relatively stable in recent years, after the increase in the early years of this millennium, caused by the migrants from the new member-states of the EU. The composition is changing in the sense that mobility to and from countries of the EU-15 is somewhat declining, but this is compensated especially by return migration from citizens of the accession countries.

The accession countries themselves do experience relatively high levels of mobility. First of all, this is a consequence of the accession itself, when it became easier for their citizens to become mobile. Especially mobility towards these countries is recently increasing as a consequence of return migration.

Finally, countries in the EU-15 that have been hit the hardest by the financial crises. Until 2008 the patterns were comparable to the other countries. Since then, they have experienced relatively high levels of mobility, consisting of two groups. First immigrants from earlier years are leaving. This even to a greater extent than it is happening in other EU-15 countries. Next to that, mobility is rising because of the emigration of nationals. Although it is not possible to prove it without any doubt, it may be supposed that this flow, to a larger extent is directed to other EU-15 countries.

4. Conclusions

This paper has tried to shed some light on the development of (im) mobility in the EU in the first years of the millennium. Mobility is and has been one of the key areas of the European Union, as it is supposed to play a crucial role in the policies aiming for cohesion within the territory of the Union. As such, in the previous millennium mobility was always considered to be too low, certainly when compared to the levels witnessed in, for instance, the US. Recent data shows that these levels have risen, but it remains questionable whether this is an effect of active policy of the European Commission and the member states, or that it is very much induced by the recent historic events, the accession of 13 new member states and the Euro, and the consequent financial crisis. It has not been the intention of this paper to determine a causal relation in this respect. However, by analysing the recent developments, what has been aimed for is to determine the plausible effect of the crises on (im)mobility in the EU.

In general, it can be seen that mobility increased in the first years of this millennium and, in a way, it seems to have functioned as an instrument of equilibrium. There are also indications that the increase has been levelling off since 2008, the year that the financial crisis peaked. Also it was suggested that the composition of the mobility flows have changed in recent years. As was shown for the exemplary countries, Denmark, Lithuania and Spain, recently some of the new accession countries have become more important as destinations for emigrants. At the same time, the share of citizens from those countries in the total emigration flows is also rising, implying that return migration has become more important in recent years. We might suggest that this is (at least partly) a consequence of the crisis. Another effect of the crisis is an increase of national emigration from economically hard hit countries. The numbers of, for instance, Spanish citizens emigrating from Spain have doubled in the last few years. In countries were the effects of the crisis are more moderate, mobility is much more stable in the recent years.

Another aim of this paper was to put the observed developments in the perspective of the threshold approach, which tries to understand individual decision with regard to mobility. Concerning the first threshold, the individual disposition towards becoming mobile, it seems that the crisis has pushed many across this hurdle. As can be seen from the Spanish case, apparently many more are considering mobility as a way out of the personal crisis. The location threshold is also very much at work in two respects. First of all cross-border mobility is still to a large extent a phenomenon that plays out over relatively short distances. Neighbouring countries do play an important role as destinations. Secondly, if mobility is taking place over longer distances, it turns out to be focused on a few countries. Spain and Italy are, for instance, important destinations for migrants from Romania. Given the dominance of a few countries, apparently there is a certain level of indifference towards many other countries in that they are less likely to go there. The trajectory threshold is less important to understand mobility patterns in the EU. This is supposed to play a bigger role in long distance and often undocumented migration. However, there is some (albeit circumstantial) evidence that it might reinforce the locational threshold. For instance, the relatively large share of Polish citizens in the Netherlands can be (partly) attributed to employment agencies that facilitated a smooth travelling back and forth between Poland and the Netherlands for seasonal workers in the agricultural sector.

Finally, it is always hard to determine whether policies have been successful. With regard to mobility, there is evidence for a modestly favourable answer in case of the EU. In the early days of this millennium, mobility increased, not only because of the expected accession of Central European countries. In recent years, there seems to be a more crisis-induced dynamic. First of all, the level of return migration is increasing, but also emigration from EU-15 countries that are strongly affected is growing. Of course it is difficult to determine whether this would also have been the case without the efforts of the EU.

Part II

Cross-Border Cooperation in Raia Ibérica

The Role of the INTERREG in Building Territorial Networks

Eduardo MEDEIROS

Centro de Estudos Geográficos, Instituto de Geografia e Ordenamento do Território – Universidade de Lisboa

1. Introduction

When it comes to the establishment of formal cross-border relations in Raia Ibérica one thing seems clear: there is a pre-INTERREG-A and a post-INTERREG-A phase. These relations were mostly informal until the establishment of INTERREG-A, in 1990. Suffice to say that the creation and establishment of cross-border networks in Raia Ibérica was provoked by this EU Community Initiative.

Consequently, it comes loud and clear that, after more than two decades of intense cross-border collaboration in Raia Ibérica, financed by INTERREG-A funds, regional and local actors saw the need to set-up long-standing cross-border structures, in order to better implement cross-border development strategies, and reinforce cross-border networks. Thus, the aim of this paper is to offer a panoramic and synthetic view of these recent (since 1990) transformations in the process of cross-border networking in Raia Ibérica, and to understand the role of the INTERREG-A Community Initiative in this process.

For the sake of the argument, this analysis is divided in three main topics. Firstly, we look upon the set-up of cross-border structures in Europe, in order to understand the importance of border regions in Europe, and also to compare Raia Ibérica with the remaining European border regions, when it comes to the establishment of cross-border structures, like the Working Communities, Euroregions, and European Groupings of Territorial Cooperation (EGTC).

The second topic provides a deeper overview of these cross-border structures in Raia Ibérica, regarding their location, strategies and the results obtained so far. This information was gathered in the cross-border structure's websites, and by interviewing some of their members. Moreover, this analysis used the information from the approved INTERREG-A projects in order to detect the main cross-border flows, and the type of entities involved in Raia Ibérica cross-border collaboration, in order to better detect the prevailing types of cross-border physical networks, proposed in the first topic.

Finally, the last topic is dedicated to analysing the strategies of the Raia Ibérica Euroregions and EGTCs, in the light of a proposed cross-border planning rationale, which invokes the need for cross-border regions to focus their strategy on two main goals: (i) barrier effect reduction and (ii) territorial capital valorisation.

2. Cross-Border Networks in Europe: the Role of INTERREG

In order to work towards a balanced and sustainable development of the EU Territory, the EU Cohesion Policy saw its beginning on 24 June 1988 with a regulation which put EU funds into the context of economic and social cohesion[1]. A couple of years later, the European Commission decided to launch a specific initiative for border regions known as INTERREG, with the central aim of promoting cross-border cooperation in economic development and to prepare the border regions to make the most of the opportunities and challenges of increasing European integration[2].

Since then, the INTERREG-A Community Initiative has maintained a central role in financing the cross-border process in Europe, and saw its intervention area spread across all the EU border areas (NUTS III), as the four programming cycles went on. The last programme (INTERREG-IV) has now become the third objective of the EU Cohesion Policy: European Territorial Cooperation[3]. Yet, despite covering almost 60% of the EU (27) area, and close to 40% of the EU population, cross-border initiatives receive less than 2% of the EU Cohesion Policy total funds (Figure 1).

1 EC, *Inforegio – Panorama – EU Cohesion Policy 1988-2008: Investing in Europe's Future*, DG REGIO, EU, Brussels, 2008.

2 EC, *Cross-Border Cooperation. Commission of the European Communities*, Directorate General for Regional Policies, Luxemburg, 1994.

3 EC, *Regions as Partners. The European Territorial Cooperation Objective*, DG REGIO, EU, Brussels, 2007.

On balance, the successive EU Cohesion Reports[4] conclude that INTERREG had a very positive role in: (i) improving interregional and cross-border relations, (ii) promoting integration and reducing the economic and social fragmentation created by borders, (iii) promoting cooperation and networking; (iv) establishing and strengthening a culture of cross-border cooperation both inside the EU and between the EU and neighbouring countries; (v) increasing the permeability of borders, both physically and administratively, (vi) exchanging experience, knowledge and know-how across borders and countries, and (vii) supporting regional development (job and business creation and assistance).

In addition, the latest EU Cohesion Report highlights the INTERREG III-A role in assisting the "creation of nearly 12,000 networks, leading to some 1,285 separate plans being formulated dealing with cross-border or transnational issues and almost 63,000 agreements being concluded"[5]. This means that 'territorial networking' has been a clear added value of INTERREG. In this regard, the rise of cross-border structures (Working Communities, Euroregions, Macro-Regions, and EGTCs) is evident in the European Space, since the implementation of the first INTERREG Community Initiative[6].

---

4 EC, *First Cohesion Report*, EU, Luxemburg, 1996; EC, *Second Report on Economic and Social Cohesion*, EU, Luxemburg, 2001; EC, *A New Partnership for Cohesion Convergence Competitiveness Cooperation*. Third report on economic and social cohesion, EU, Luxemburg, 2004; EC, *Growing Regions, Growing Europe*, Fourth report on economic and social cohesion, EU, Luxemburg, 2007; EC, *Investing in Europe's Future*, Fifth report on economic, social and territorial cohesion, EU, Luxemburg, 2010.

5 EC, *Investing in Europe's Future*, *op. cit.*, p. 221.

6 See E. Medeiros, "Old vs Recent Cross-Border Cooperation: Portugal-Spain and Sweden-Norway", in *AREA*, Vol. 42, No. 4, 2010, pp. 434-443; E. Medeiros, "(Re) defining the Concept of Euroregion", in *European Planning Studies*, Vol. 19, No. 1, 2011, pp. 141-158; E. Medeiros, "Euro-Meso-Macro: The New Regions in Iberian and European Space", in *Regional Studies*, published online: http://www.tandfonline.com/doi/abs/10.1080/00343404.2011.602336, 2011; M. Perkmann, "Cross-Border Regions in Europe – Significance and Drivers of Regional Cross-Border Co-operation", in *European and Urban Studies*, Vol. 10, No. 2, 2003, pp. 153-171; C. Huggins, "Motivations behind Local Government Transnational Networking", in *Regional Insights*, Vol. 4, No. 1, 2013, pp. 9-13.

Figure 1. *INTERREG-A Data.*

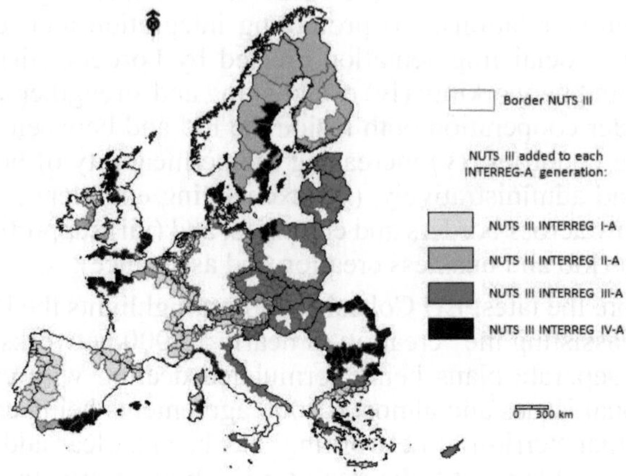

INTERREG-A	1989-1993	1994-1999	2000-2006	2007-2013	EU-27
Area (1000 km ^{2})	559	1,283	1,925	2,495	4,234
Area (% EU 27)	*13.2*	*30.3*	*45.5*	*58.9*	*100*
GNP – (€ per capita) – 2006	24,400	25,038	20,764	20,581	25,900
Population (millions) – 2006	64	95	156	204	493
Population (% EU 27) – 2006	*13.0*	*19.3*	*31.6*	*41.4*	*100*
Pop. Density (inhab/km^2) – 2003	260	235	230	276	113
Crude Death Rate (‰) – 2003	10.8	10.4	10.4	10.5	9.6
Crude Birth Rate (‰) – 2003	10.1	9.6	9.2	9.3	10.3
Allocation funds (million €)	1,034	2,617	3,998	8,723	
Allocation funds (% EU Structural Funds)	*0.7*	*1.3*	*1.8*	*1.8*	
Programmes	*31*	*59*	*64*	*52*	

Source: Data (EC reports + Eurostat) – Author compilation.

Curiously, the older western European cross-border Working Communities and some Euroregions have spread out to Eastern Union, with the Euroregion designation. However, most often, they are far from being a genuine and intense form of cross-border cooperation[7]. In later years, and more specifically, since 2008, the possibility of establishing European Groupings of Territorial Cooperation (EGTCs), as legal cross-border entities with associated powers and obligations, has seen the set-up of some of these

[7] See E. Medeiros, "(Re)defining the concept of Euroregion", *op. cit.*, pp. 141-158.

cross-border structures in Europe, although at a smaller pace, than initially expected. Concerning the most recent form of cross-border entities (EGTCs), Hungary, Slovakia, France, Spain, Belgium and Portugal (Figure 2) took the leading step in implementing these innovative forms of management and implementation of territorial cooperation programmes or projects.

Figure 2. *Cross-Border Structures in Europe – June 2013.*

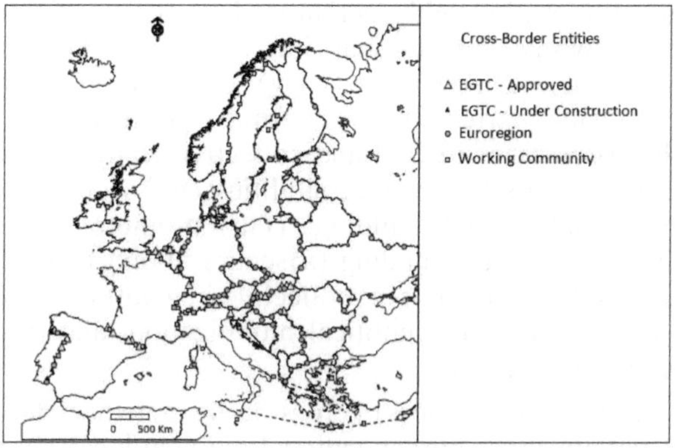

Source – Data: (Committee of the Regions and AEBR) – Author cartography.

As can be understood by the reading of the previous figures, there is a wide range of different realities concerning the EU cross-border structures: (i) some are located in urban areas, some in rural; (ii) some have been cooperating for a long-time, others for a short-time; (iii) some are based on local interactions, others are fostered on the regional arena; (iv) some have a long-standing cross-border strategy and intense contacts, others the opposite. Yet, all of them can be considered cross-border networks if they are seen as cross-border "arenas in which decision-makers and interests come together to mediate differences and search for solutions"[8].

In the geographical perspective, and following the theory of networks and flows, these cross-border networks can be seen as cross-border geographical systems, which carry "flows of people, goods, messages and anything else that is moved from one place to another"[9], within a concept

[8] N. Nugent, *The Government and Politics of the European Union*, Basingstoke, Palgrave, Macmillan, 2006, p. 574.

[9] R. Abler, J. Adams, and P. Gould (eds.), *Spatial Organization. The Geographers View of the World*, London, Prentice-Hall International, Inc, 1972, p. 255.

"that connotes a whole that is greater than the sum of its parts"[10]. Usually, in these types of networks the nodes are the urban agglomerations, while the lines are made up of roads and railways.

However, these cross-border networks are very much connected with the concept of policy networking, "designed to construct coalitions based on common policy interests"[11]. On the other hand, this process intends to: (i) strengthen a more decentralized, pluralist model of government in highly centralized countries; (ii) promote the cross-border networks institutional interests; (iii) improve regional comparative economic advantages; (iv) develop social bonds and subcultures outside structures of formal political accountability; (v) establish processes of mutual and collective learning; (vi) facilitate the construction of cross-border accessibilities; (vii) secure organisational learning and social capital building; (viii) promote local identities; (ix) compete for influence and funding; (x) secure funding for its participants; (xi) share knowledge; (xii) support collective innovations; (xiii) promote sustainable development and political mobilization; and (xiv) promote opportunities for democracy[12].

In other words, the robustness of cross-border networks depends on the presence of certain cross-border national/regional/local characteristics, and on the level, or intensity, of several relational components. In the end, the latter can be measured by the effects of the cross-border cooperation

---

10 R. Banai, and T. Wakolbinger, "A Measure of Regional Influence with the Analytic Network Process", in *Socio-Economic Planning Sciences*, Vol. 45, 2011, p. 167.

11 A. Church, and P. Reid, "Urban Power, International Networks and Competition: The Example of Cross-Border Cooperation", in *Urban Studies*, Vol. 33, No. 8, 1996, p. 1297.

12 A. Church, and P. Reid, "Urban Power, International Networks and Competition: The Example of Cross-Border Cooperation", *op. cit.*; A. Church, and P. Reid, "Cross-Border Co-operation, Institutionalization and Political Space across the English Channel", in *Regional Studies*, Vol. 33, No. 7, 1999, pp. 643-655; S. Dühr, C. Colomb, and N. Vincent, *European Spatial Planning and Territorial Cooperation*, London, Routledge, 2010; A. Faludi, *Cohesion, Coherence, Cooperation: European Spatial Planning Coming of Age?*, London, Routledge, 2010; C. Huggins, "Motivations behind Local Government Transnational Networking", *op. cit.*; P. Valkering, *et al.*, "An Analysis of Learning Interactions in a Cross-Border Network for Sustainable Urban Neighbourhood Development", in *Journal of Cleaner Production*, Article in Press, 2012, pp. 1-10; R. Rutten, and F. Boekema, "Regional Social Capital: Embeddedness, Innovation Networks and Regional Economic Development", in *Technological Forecasting & Social Change*, Vol. 74, 2007, pp. 1834-1846; J. Dlouhá, *et al.*, "Social Learning Indicators in Sustainability-oriented Regional Learning Networks", in *Journal of Cleaner Production*, Article in Press, 2012, pp. 1-10; M. Perkmann, "Construction of New Territorial Scales: A Framework and Case Study of the EUREGIO Cross-Border Region", in *Regional Studies*, Vol. 41, 2007, pp. 253-266; P. Hall, "Opportunities for Democracy in Cross-Border Regions? Lessons from the Øresund Region", in *Regional Studies*, Vol. 42, No. 3, 2008, pp. 423-435.

process in a series of territorial elements (Table 1). There are, obviously, many different types of relational and physical cross-border networks in Europe, due to the territorial variety of European border regions. Yet, when it comes to the physical characteristics of cross-border networks, the main differentiator factor is the territorial proximity factor of the cross-border process. Here, three scenarios can be found. Firstly, cross-border relations are stronger between large urban agglomerations, located at a large distance from the border. In the second scenario, proximity matters, as cross-border relations are stronger between the urban agglomerations in close proximity to the border. Lastly, the third scenario is a mix of the previous two (Figure 3).

Table 1. Cross-Border Networks Robustness Elements

Cross-Border Networks Robustness Pre-Conditions
– Existence of a common policy interest – Existence of a common cross-border strategy – Existence of cross-border structures (Working Communities, Euroregions, EGTCs) – Existence of regional, local and urban involvement – Existence of stability in institutional relations – Exchange of knowledge and information – Synchronize policy implementation
Cross-Border Networks Robustness expected effects
– Increase of monetary resources, mainly from EU programmes – Increase of economic benefits – Increase regional visibility in the international arena – Increase of mutual and collective learning – Increase in the dissemination of good practices – Increase in living conditions – Increase functional integration – Improve physical infrastructures – Reduction of core-periphery disparities – Reducing transition costs – Regional, local and urban autonomy, empowerment and decentralization – Regional, local and urban governance – Develop cross-border planning and sustainable development practices – Generate mutual trust – Secure organizational learning and social capital building – Promote urban revitalization

Figure 3. *Cross-Border Types of Physical Networks.*

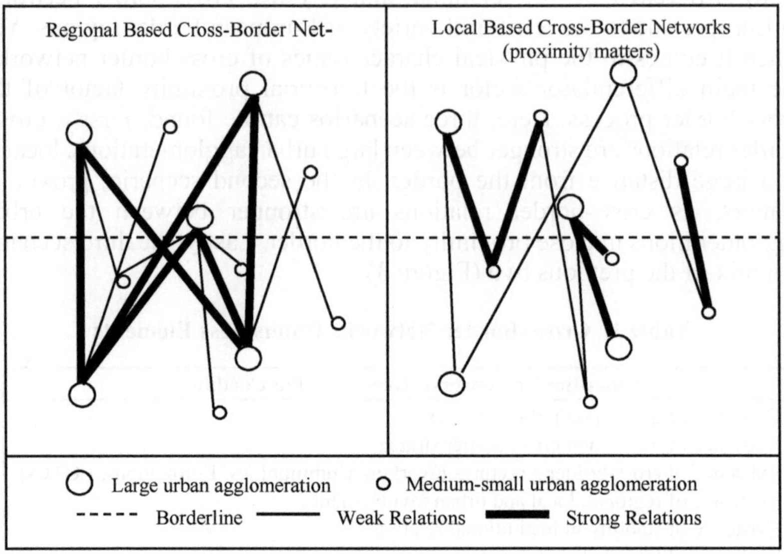

Source: based on ESPON 1.1.1, Potentials for Polycentric Development in Europe, Final Report, ESPON, Luxemburg, 2004, p. 174.

3. Cross-Border Networks in Raia Ibérica

Raia Ibérica is the term used to designate the Portuguese-Spanish Border Area. The limits of Raia Ibérica are not easily defined. Some use the border municipality's area to study it[13]. Nevertheless, in recent years, the norm has been the use of the border NUTS III to define the Raia Ibérica space, mainly due to the INTERREG-A criteria in identifying European border areas. Within these criteria, Raia Ibérica holds about 1/5 of the Iberian Peninsula area, but only 10% of the population (Table 2).

[13] See C. Cavaco, "Um olhar sobre a fronteira Portugal-Espanha", in *As regiões de fronteira – inovação e desenvolvimento na perspectiva do Mercado Único Europeu, Estudos para o desenvolvimento regional e urbano*, Lisboa, CEG, UL, 1995, pp. 19-34; L. Trigal, R. González, and F. Guichard, *La articulación territorial de la raya hispano-portuguesa*, Actas Simposium Vilar Formoso, Fundación Rei Afonso Henriques, Serie monografía y estudios, Zamora, 1996.

Table 2. Raia Ibérica Geographic and Demographic Data – 2011

Territorial Unit	Area (km^2)	% Territory	Population 2011	% Population	Density ($inhab/km^2$)
Spain	505,988	100.0	47,265,321	100	93.4
Portugal	92,152	100.0	10,562,178	100	114.6
Spain and Portugal	598,140	100.0	57,827,499	100	96.7
Spanish Border Area	86,441	17.1	3,461,852	6	40.0
Portuguese Border Area	50,574	54.9	2,106,707	20	41.6
Raia Ibérica	**137,015**	**22.9**	**5,568,559**	**10**	**40.6**

Source: Data (National Statistics) – Adapted.

In short, Raia Ibérica can be regarded as one of the less populated and developed (socioeconomically) areas in the Iberian Peninsula, with the exception of a few dynamic urban axis along the urbanized littoral coast (North/Galicia and Algarve/Andalucia)[14]. In the same line, before the enlargement of the EU with the eastern countries, this was definitely the least developed border area of the EU space. This justifies the fact that Raia Ibérica is the EU border area which has received the largest piece of cake of the INTERREG-A funds, so far.

Indeed, the INTERREG-A has been key in reducing the historically strong barrier effect which has prevailed for almost 800 years, along the unchanged boundary between Portugal and Spain, with minor exceptions in the North-Galicia border area. Yet, the first two INTERREG-A generations (1989-1993 and 1994-1999) produced few genuine cross-border projects. Nevertheless, they laid the foundations of the cross-border process, and witnessed the set-up of the first cross-border associations, which have reached a substantial number (more than 80)[15] in recent years. In this regard, a significant emphasis should be put in the set-up of three Euroregions and three EGTCs, while others are being prepared (Figure 4).

14 See E. Medeiros, *A Cooperação Transfronteiriça Portugal-Espanha e Suécia-Noruega. Efeito Barreira e Impactes Territoriais do INTERREG-A*, Tese de Doutoramento, IGOT, Lisbon University, 2010.

15 See E. Medeiros, *A Cooperação Transfronteiriça Portugal-Espanha e Suécia-Noruega. Efeito Barreira e Impactes Territoriais do INTERREG-A*, *op. cit.*

Figure 4. *MEGAS and FUAS / Iberian Euroregions and EGTC – 2013.*

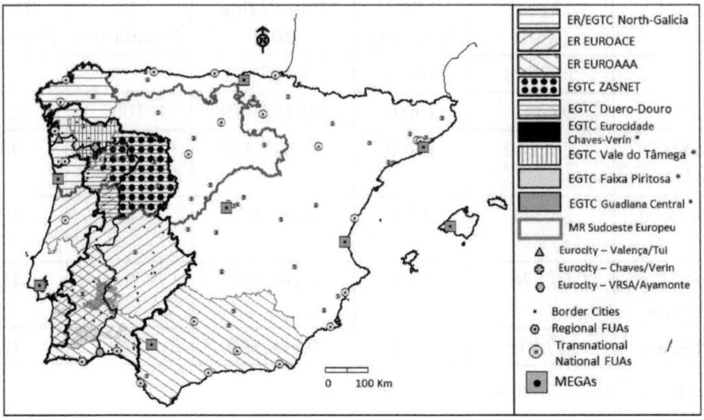

Source: Data (several). Author cartography.

To gain a clear understanding of the results produced so far by these cross-border structures, we interviewed some of their members. However, not all the representatives of these cross-border structures manifested their openness to speak about the results obtained so far. This is definitely a negative sign. Even so, the answers provided by the three respondents (EUROACE, EUROAAA and EGTC Duero-Douro) were mostly positive, when it comes to the experience in promoting cross-border cooperation, so far. In short, the following positive notes were highlighted:

- Higher inter-institutional integration;
- Higher quality and quantity in the process of cross-border cooperation (CBC);
- Higher analysis and intervention capacity;
- Higher possibilities to develop ideas and actions;
- Higher possibilities to guarantee the continuity of the CBC process;
- Higher possibilities in producing an integrated strategy;
- Higher possibilities in accessing flagship cross-border projects;
- The voice of the region is heard, unlike in previous times (participation in international forums – like the OpenDays);
- The territory is seen as one;
- The responsibility of the projects is shared;
- The governance is now more participated;
- The bases for a new future were established.

On the negative side is the mistrust of the regional and national entities towards these cross-border structures. For instance, these administrative levels do not want to give management capacity to these cross-border spaces. In a sense, the spirit of the set-up EGTCs is not being put into place (based on the subsidiarity principle), when it comes to letting the border areas delineate their development strategies. On the other hand, the regional asymmetries are still present, regarding the territorial dimension, which is larger in the Spanish side. Furthermore, on that side of the border, there is a regional administrative body with political competences, unlike what happens in the Portuguese side.

Be that as it may, there seems to be a growing interest in establishing new EGTCs in Raia Ibérica, mainly in areas further away from the main border urban settlements, as a clear sign of the need to reinforce local empowerment. These trends might have some logic, due to the highly centralized form of government in Portugal, which lacks the regional administrative level. In turn, the local (municipalities) administrative level is quite dynamic and has a vast number of administrative competences. In contrast, in Spain the local level has fewer administrative competences, while the regional administrative level has administrative capacity to take their own decisions, in many development areas, without consulting Madrid.

Certainly, these differences in the governance process in both Iberian countries justifies the prevailing cross-border flows in the Raia Ibérica (Figure 5), which are also dependent on the design of the INTERREG-A Programme, which is divided in five sub-regions (1 – Galicia/Norte de Portugal; 2 – Castilla y León/Norte de Portugal; 3 – Castilla y León/ Centro de Portugal; 4 – Extremadura/Centro/Alentejo; 5 – Andalucía/ Alentejo/Algarve – Figure 6).

A deeper look at the cross-border flows (INTERREG-A projects) in Raia Ibérica shows a prevailing 'Regional Based' type of Cross-Border Networks. As a matter-of-fact, the stronger cross-border links are established between the main regional urban agglomerations, even if they are located outside Raia Ibérica (Oporto -P, Santiago de Compostela- S, Valladollid- S, Coimbra- P and Seville- S). Why? Because these cities, together with Zamora (S), Salamanca (S), Évora (P), Merida (S) and Faro (P), are the regional administrative centres of Raia Ibérica (where the main administrative decisions are made).

Figure 5. *INTERREG-A (III+IV) Main Networks and Cross-Border Flows.*

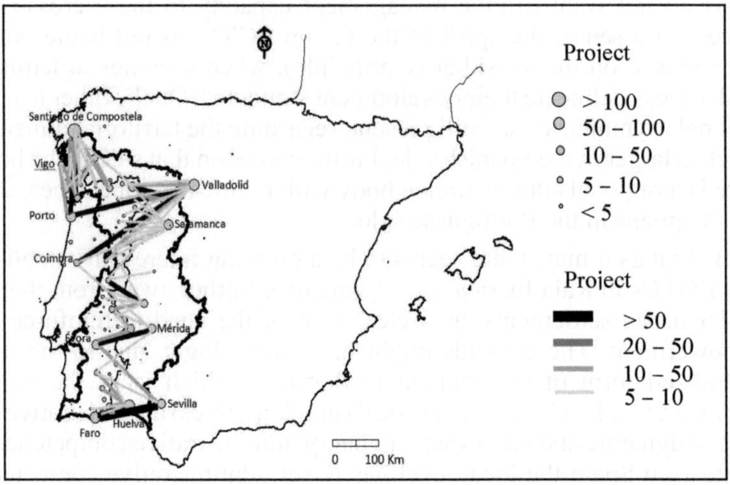

Figure 6. *Portuguese – Spanish – INTERREG-A Sub-programmes.*

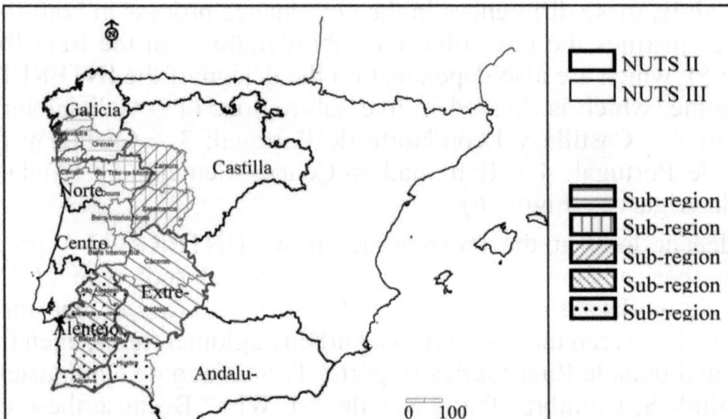

This does not mean that medium-strong flows do not take place between smaller urban centres along the border. This is especially true in the north of Portugal (Minho) – Galicia border area, and in several other cross-border cooperation 'hot spots' located very close to the border (Chaves-Ourense, Bragança-Zamora; Guarda-Salamanca; Castelo Branco-Caceres and Faro-Huelva). Hence, on certain situations, there is a mix of both regional and local based cross-border networks in Raia Ibérica.

Going deeper in this analysis, it is curious to see that the prevailing Regional domain in the cross-border cooperation process in Raia Ibérica is also quite clear when we look upon the type of actors which play the major role in this process (Figure 7). Here, the public institutes lead the way, although they have been losing ground in the more recent INTERREG-A generations. Needless to say, these institutes are located, mostly, in the regional capitals, as the remaining regional entities and universities (Figure 8). Concerning the latter, it is fair to say that the present INTERREG-A generation has, finally, given this type of entities a prominent role, contrary to the previous one.

Finally, the role of the municipalities in the cross-border cooperation process in Raia Ibérica has been of vital importance, and has not diminished over time. For one, the lack of regional autonomy in Portugal explains this presence (Figure 9). Secondly, the possibility of a strong presence of the Municipalities in the EGTCs has also driven new interest in attracting the local authorities to the cross-border arena, in Raia Ibérica.

Figure 7. *Portuguese – Spanish INTERREG-A Type of Actors (1990-2013)*.

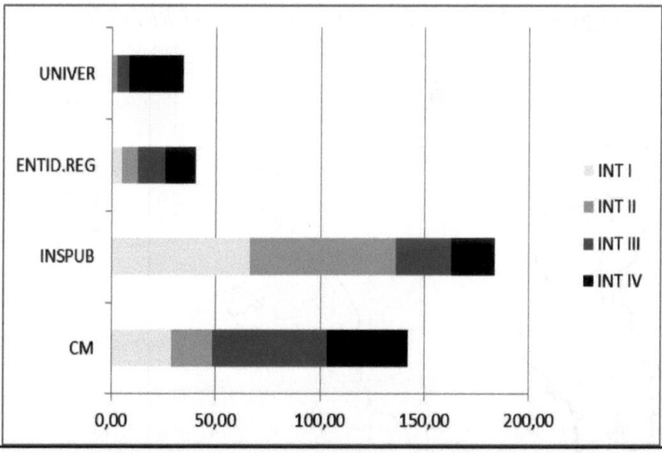

UNIVER – Universities; REGENT – Regional Entities; PUBINT – Public Institutes; MUNICIP – Municipalities

Figure 8. *Universities' Presence in INTERREG III-A.*

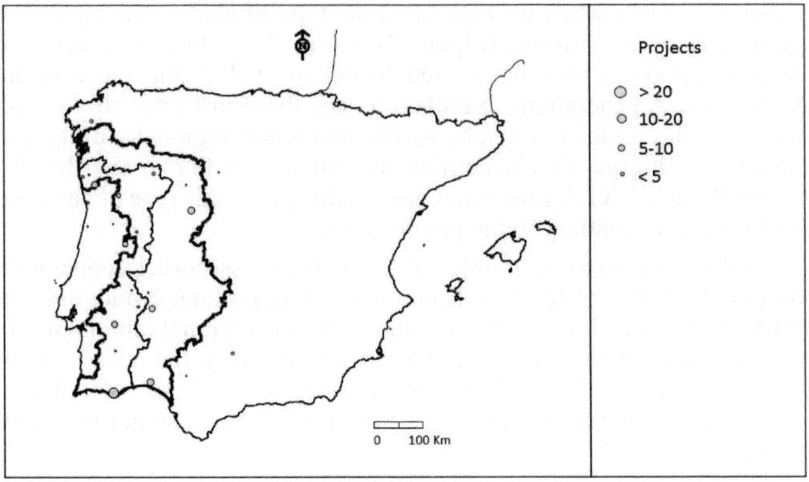

Source: Data (several). Author cartography.

Figure 9. *Municipality's Presence in INTERREG III-A.*

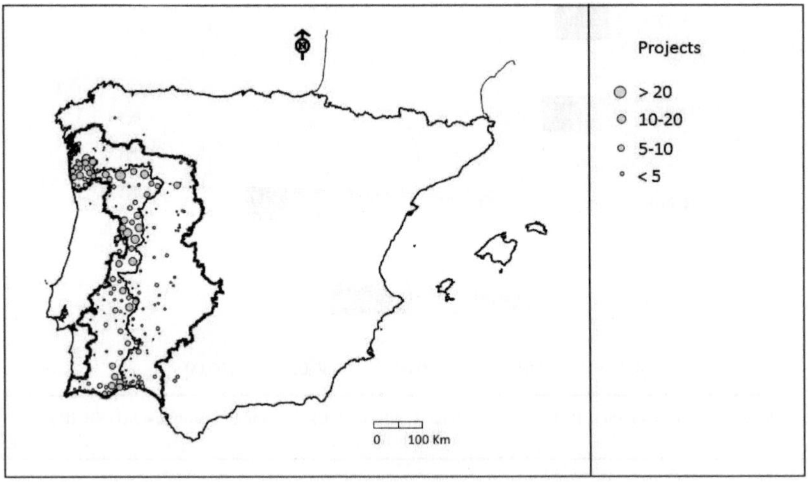

Source: Data (several). Author cartography.

4. New Cross-Border Strategies in Raia Ibérica?

From the European Spatial Development Perspective[16], to the recent Territorial Agenda 2020[17], the process of territorial cooperation in Europe has been recognized as key to foster the EU regional development. In concrete terms, the former document highlights the need to promote strategic partnerships and co-operation between urban regions, which require a regional, cross-border and transnational approach[18].

Later on, the first EU Territorial Agenda[19] considers territorial cooperation as a vital aspect to strengthen European networks of cities and regions. One year later, the Green Paper on Territorial Cohesion[20] advanced three main goals to achieve Territorial Cohesion in Europe: (i) Concentration (to overcome differences in density); (ii) Connecting Territories (to overcome division) and (iii) Cooperation (to overcome division). Finally, the latest Territorial Agenda[21] highlights the benefits of territorial cooperation in establishing Euroregions or twin city links.

Territorial cooperation and, in particular, cross-border cooperation is, nevertheless, a complex issue and requires spot on and adequate strategies to make the most out of the available funds while being implemented. In this regard, a mix of top-down (the regional view) and bottom-up (the local view) approaches should complement themselves in order to focus the design of cross-border strategies in two main dimensions: (i) reduce the barrier effect along the border, in all its dimensions, and (ii) valorise the territorial capital of the border region (Figure 10).

Contrary to this, there is a prevalent tendency for each side of the border to look at their own regional development interests, and to use the

[16] EC, *European Spatial Development Perspective. Towards Balanced and Sustainable Development of the Territory of the European Union*, European Communities, Luxemburg, 1999.

[17] EC, *Territorial Agenda of the European Union 2020. Towards an Inclusive, Smart and Sustainable Europe of Diverse Regions*, agreed at the Informal Ministerial Meeting of Ministers responsible for Spatial Planning and Territorial Development, 19th May 2011, Hungary, Gödöllő, 2011.

[18] EC, *European Spatial Development Perspective. Towards Balanced and Sustainable Development of the Territory of the European Union, op. cit.*

[19] EC, *Territorial Agenda of the European Union. Towards a More Competitive and Sustainable Europe of Diverse Regions*, 24th/25th May, Leipzig, 2007.

[20] EC, *Green Paper on Territorial Cohesion. Turning Territorial Diversity into Strength*, Communication from the Commission to the Council, the European Parliament, the Committee of the Regions and the European Economic and Social Committee, October 2008, Brussels, 2008.

[21] EC, *Territorial Agenda of the European Union 2020, op. cit.*

available EU funds from cross-border cooperation for their own purposes. This was very clear in the first two Portuguese-Spanish INTERREG-A generations. In the third, however, this unilateral rational of implementing cross-border projects was less active, due to regulation demands in forcing genuine cross-border processes. Even so, overall, the idea that cross-border cooperation strategies should be focused on supporting projects which reduce the barrier effect is still far from being generally accepted, in the European space.

However, one of the big advantages in establishing cross-border entities is the possibility to solidify a more integrated vision to the territorial development of the border area. Yet, if we look at the main strategies of the three Raia Ibérica Euroregions (Table 4) and the three Raia Ibérica EGTCs (Table 5), we can clearly see that, for the most part, they follow the rationale of a typical regional development strategy. Not unexpectedly, these strategies try to make the most of the border area territorial potentials, and do not put particular focus in the barrier effect reduction. Moreover, they are somewhat vague and unrealistic. Vague because they propose too general goals, instead of choosing three or four specific goals, to use the available funds. Unrealistic because the available funds from the INTERREG-A have proven insufficient to invert the territorial exclusion path which most European border areas faced in the last couple of decades[22].

[22] E. Medeiros, *A Cooperação Transfronteiriça Portugal-Espanha e Suécia-Noruega. Efeito Barreira e Impactes Territoriais do INTERREG-A*, *op. cit.*; ESPON, *Cross-Border Cooperation. Cross-Thematic Study of INTERREG and ESPON activities*, INTERACT, ESPON, Luxemburg, 2007.

Figure 10. *Cross-Border Planning Main Dimensions and Components.*

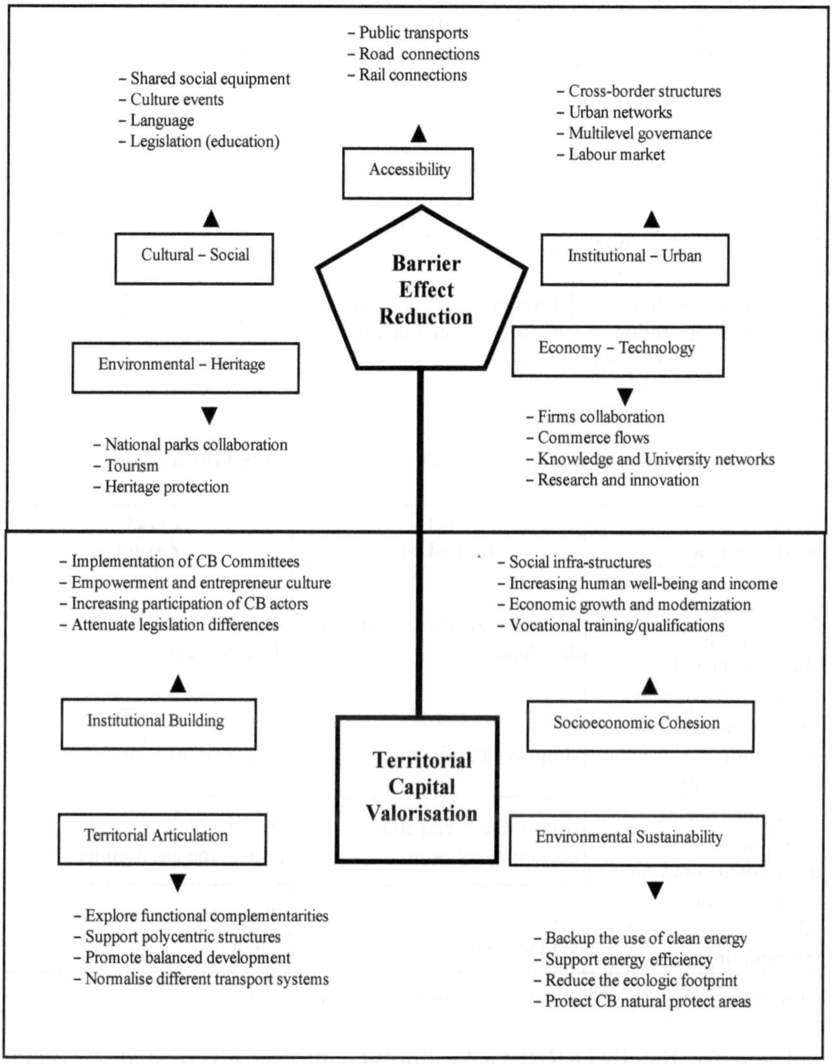

Source: Medeiros, 2013[23].

[23] E. Medeiros, "Is there a new TRUST in Inner Scandinavia", in *Geografiska Annaler* (being reviewed): http://ww3.fl.ul.pt//pessoais/Eduardo_Medeiros/docs/PUB_PAP_EM_Is_there_a_new_trust_in_IS.pdf [accessed on 9th July 2013].

Table 3. Raia Ibérica Euroregions Main Strategic Guidelines

North-Galicia (EURONORGAL)	Alentejo-Centro-Extremadura (EUROACE)	Alentejo-Algarve-Andalusia (EUROAAA)
-Promoting competitiveness and employment promotion	-Promoting competitiveness and employment	-Promoting competitiveness and employment
-Environment, heritage and risk prevention	-Environment, heritage and natural environment	-Environment, heritage and natural environment
-Planning and accessibility	-Accessibility and spatial planning	-Accessibility and spatial planning
-Socio-economic and institutional integration	-Promotion of cooperation and socioeconomic integration	-Promotion of cooperation and socioeconomic integration

Source: Medeiros, 2011[24].

Table 4. Raia Ibérica EGTC Main Strategic Guidelines

EGTC North-Galicia	EGTC Duero-Douro	EGTC Zasnet
-Promote economic development and competitiveness – Innovation and knowledge exchange	-Promote local and economic development	-Promote economic development
-Improve attractiveness – transportation systems and accessibility	-Improve public transports	-Improve infra-structures
-Strengthen social, economic and institutional cohesion	-Reinforce social and economic cohesion	-Promote tourism, culture and protect the environment
-Establish more efficient instruments for programming	-Facilitate cross-border cooperation	-Promote cross-border relations

Source: Data (Committee of the Regions) – Author elaboration.

Even so, the INTERREG-A approved projects have had a crucial role in reducing the barrier effect along the Raia Ibérica boundary[25].

---

24 E. Medeiros, "Euro-Meso-Macro: The New Regions in Iberian and European Space", in *Regional Studies*, published online: http://www.tandfonline.com/doi/abs/10.1080/00343404.2011.602336, 2011.

25 E. Medeiros, *Efeito Barreira e Cooperação Transfronteiriça na Raia Ibérica*, Iberografias No. 18, Centro de Estudos Ibéricos, Lisbon, Âncora Editora, 2011.

In this regard, the permeability of the border was largely increased in the accessibilities domain, mainly during the first two INTERREG-A generations (Figure 11). These investments favoured mostly the North-Galician and Algarve-Andalucia sub-regions (Figure 12). In market contrast, the two most recent INTERREG-A programming cycles have put more emphasis on the reduction of the economy-technology and the environment-heritage dimensions, which is a sign of the increasing maturity of the Portuguese Spanish INTERREG-A programme.

Figure 11. *Portuguese-Spanish INTERREG-A Investment per Barrier-Effect Dimension (1990-2013).*

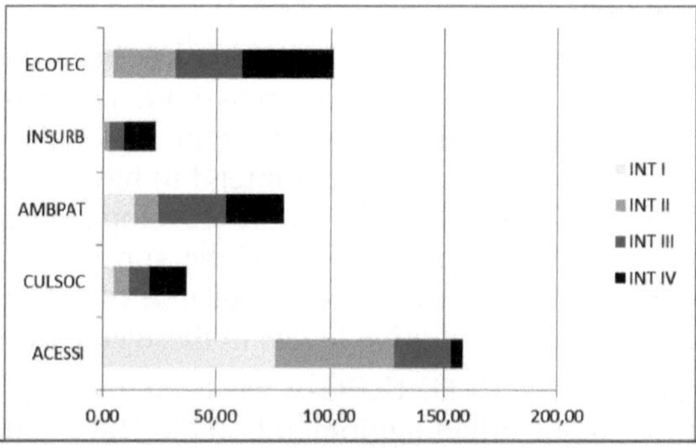

ECOTEC – Economy-Technology; INSURB – Institutional-Urban; ENVHER – Environment-Heritage; CULSOC – Cultural-Social: ACCESS – Accessibilities

Figure 12. *Portuguese-Spanish INTERREG-A Investment per NUTS III (1990-2013).*

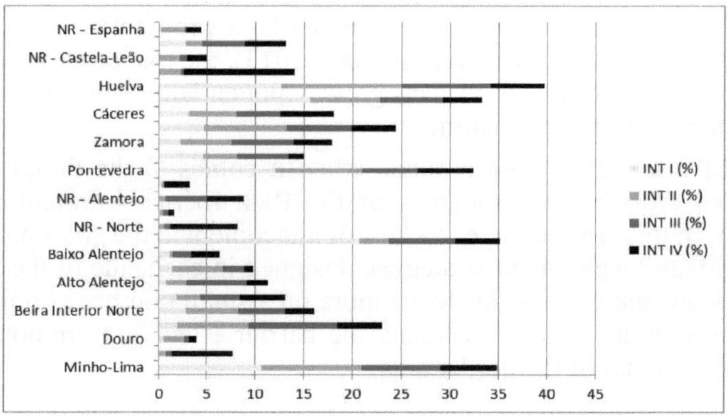

5. Conclusions

Over the past two decades, the INTERREG-A EU Community Initiative has played a crucial role in reducing the barrier effect along the Raia Ibérica border. Alongside, it ignited the set-up of several cross-border structures (Working Communities, Euroregions, Eurocities and EGTCs), along the border area.

Indeed, at present times, Raia Ibérica can be regarded as one of the most dynamic cross-border territories in Europe, regarding the implementation of such cross-border structures, which are crucial in sustaining collaborative networks. This can be seen in the pioneering set up of three EGTCs (with several other on the pipeline), three Eurocities, and three Euroregions.

Here the north-eastern part of Raia Ibérica has been the most active one in promoting the establishment of cross-border structures; perhaps because this border area is further away from regional administrative centres, and some local political leaders intend to have a more proactive and independent role in deciding the future development of the region. Yet, according to some, the application of the subsidiarity principle is far from being achieved in these regions, because of the mistrust of the regional and national administrative levels in these endeavours.

On the other hand, the INTERREG-A has been giving greater priority to connecting the major urban agglomerations in Raia Ibérica, which is justified by the presence of the main regional administrative offices in these places. Indeed, the regional institutes were the dominant type of actors in managing INTERREG-A projects, during all the four cycles (1989-2013), and especially in the first two. Nevertheless, the lack of the regional administrative level in Portugal, and the subsequent powerful role of the municipalities, justifies their large presence in promoting cross-border partnerships within the INTERREG-A programme. Yet, in the present INTERREG-A generation (2007-2013), the role of universities in managing INTERREG-A projects grew exponentially, which is a sign of the maturity of this programme.

Finally, a critical note is made when it comes to the design of the proposed cross-border strategies of the Raia Ibérica Euroregions and EGTCs. In this regard, there is a lack of attachment to the goals present in the regional development strategies, designed in each side of the border. In our view, these strategies focus more on a small number of objectives with the ultimate goal of reducing the barrier effect and promoting the territorial capital of the border area.

An Analysis of Slovenian Cross-Border Cooperation Programmes with Austria and Italy

Alessandro TORELLO

University of Siena

1. Introduction

The purpose of this chapter is to present a concise analysis of Slovenian cross-border cooperation initiatives with Austria and Italy, according to Cross-Border Cooperation Operational Programmes. The chapter is divided into three parts.

Part one focuses on the cooperation between Slovenia and Austria, since the implementation of Interreg III A Programme (2000-2006), as this programme was the basis for the subsequent launch of the 2007-2013 and the forthcoming 2014-2020 programmes.

Part two analyses the current cross-border cooperation programme (2007-2013) which has been implemented in order to reduce economic and social disparities between Slovenia and Italy.

Finally, part three pays attention to the enhancement of transport networks in Slovenia, in relation to sustainable development standards and the freedom of movement of citizens, workers and goods.

2. Cross-Border Cooperation Programmes between Austria and Slovenia

2.1. Interreg III A Programme (2000-2006)

The Interreg III A Programme for cooperation between Austria and Slovenia was launched in 2000, four years before the accession of Slovenia to the European Union.

The Programme had the purpose of guaranteeing the social, economic and rural development of Austro-Slovenian border regions[1].

[1] Interreg III A (Austria – Slovenia Cross-Border Cooperation) was based on Commission decision C (2001) 2043 of 6th August 2001. Decision C (2001) 2043 was revised by

The total budget issued for the programme was over 63 million euro, of which 33.5 million euro was provided by EU Structural Funds. The 2000-2006 Interreg III A plan was based on five priorities, which were considered imperative for the subsequent 2007-2013 programme:

Priority 1. Economic cooperation, in particular on the following fields: Business start-ups, tourism and sustainable agricultural production.

Priority 2. Human resources and regional cooperation investments, which focused on the need to dismantle linguistic and cultural barriers through cultural exchange and encounter programmes, i.e. supporting projects relevant to the cooperation of cities, schools and cultural associations. It was also considered fundamental to enhance investment in human resources and regional cooperation in order to implement a common labour market.

Priority 3. Sustainable spatial development, whose goals were: Protecting the natural environment and conserving biodiversity (the endangered ecosystem of Alpine areas); testing the quality of waters of the cross-border fluvial system (Drau/Drava and Mur/Mura); ensuring sustainable development (use of renewable energy resources). Also Priority 3 focused on the improvement of transport infrastructures and public transport.

Priority 4. General support for border regions. The majority of the funds for Priority 4 were spent to help Slovenia access the European single market progressively.

Finally, Priority 5 was about technical assistance support.

The implementation of Interreg III A faced several difficulties due to evident disparities between Austrian regions and Slovenian border areas. First of all, during the 1990s and at the beginning of the new century, Slovenian border regions were characterized by an old heavy industry and an inefficient market economy, as Yugoslavia's previous planned economy experience still had some impact on the market. On the contrary, Austria, which joined the EU in 1995, presented regions with a vital and dynamic market economy (e.g. Styria and Carinthia). Secondly, the Slovenian GDP per capita in the border area was nearly 65% of the average of the EU, while in Austrian regions GDP per capita was about 130% of the average of the EU.

---

the following Decisions: Commission Decision C (2002) 1703 of 26th July 2002; Commission Decision C (2004) 4154 of 19th October 2004; Commission Decision K (2005) 4971 of 5th December 2005. For more details concerning Interreg III A (Austria – Slovenia Cross-Border Cooperation), see also http://ec.europa.eu/regional_policy/country/prordn/details.cfm?gv_PAY=SI&gv_reg=ALL&gv_PGM=241&LAN=7&gv_PER=1&gv_defL=7 [accessed on 5th July 2013. This site was archived on 15th April 2011, yet statistical information is still available].

2.2. Key Elements of the 2007-2013 Cooperation Operational Programme

The European Commission approved a Cross-Border Cooperation Operational Programme between Slovenia and Austria for the period 2007-2013 in December 2007[2]. This operational programme was elaborated in accordance with Regulation of the Council 1083/2006 of 11 July 2006 relevant to the European Regional Development Fund (ERDF), the European Social Fund (ESF), the Cohesion Fund, and the Regulation of the European Parliament 1080/2006 of 5 July 2006 concerning the European Regional Development Fund. The Programme, through the European Regional Development Fund, intended to encourage cooperation and economic integration between the Austrian and Slovenian border regions, taking into consideration the different levels of development of every region. The programme area stretches from the Alpine mountain regions in the West and in the North to the Burgenland-Styrian-Slovene border to the East. It also includes part of the Pomursko region, part of Podravsko, and the region of Südburgenland up to the Pannonian plain. The area has two cross-border rivers: the Mura (Mur) and the Drava (Drau)[3].

The length of the Slovene-Austrian border is 330 km, 170 km of which on the Austrian side belong to Kärnten, 145 km to Steiermark, and only 15 km to Burgenland. The Programme covers 35,523 square kilometers of territory (23,899 square kilometers in Austria and 11,625 in Slovenia). The total population involved in the Programme is more than 3,200,000 people. Concerning Slovenia, the Programme covers around 57% of its territory and involves more than 70% of the whole population. The core areas are represented by the capitals of the regions and the main industrial centres: Kranj, Ljubljana, Celje, Maribor in Slovenia; Graz, Villach and Klagenfurt in Austria[4].

The Austrian industrial areas of the Programme are Obersteiermark Ost, Obersteiermark West and Unterkärnten – over 35% of employees work in the industrial sector. The service sector is primarily in Graz,

---

2 European Commission, *European Territorial Cooperation 2007-2013: Cross-Border Cooperation between Austria and Slovenia – Operational Programme*, CCI number: 2007CB163PO054.

3 The territorial areas involved in the project on the Austrian side are: Oststeiermark West, Südsteiermark, Klagenfurt, Villach, Unterkärnten and Südburgenland. On the Slovenian side: Gorenjsko, Koroško, Savinjsko, Podravsko and Pomursko.

4 European Commission, *European Territorial Cooperation 2007-2013: Cross-Border Cooperation between Austria and Slovenia – Operational Programme*, *op. cit.*, p. 10. Source: Östat (http://www.statistik.at) [accessed on 10th May 2013]; Statistični urad Republike Slovenije (Statistical Office of the Republic of Slovenia), http://www.stat.si [accessed on 2nd July 2013].

Klagenfurt and Villach (over 77% of employment). Without considering farmers, Austrian rural areas are marginal in terms of employment shares (nearly 1-2%). In Slovenia, the most developed industrial regions of the Programme are Gorenjsko, Koroško, Savinjsko and Podravsko. Industries are also present around the main urban centres (Ljubljana, Maribor and Celje). An important rural region of the Programme is Pomursko, where 14% of employees work in agriculture. The majority of the population in Osrednjeslovensko work in various service sectors (69.7%). However, the growth of the service sector has been considerable in each region recently, especially in Osrednjeslovensko and Pomursko[5].

The accession of Slovenia to the European Union in May 2004 allowed the country to cooperate with Austria in specific economic branches, such as the labour market and employment opportunities, competitiveness and development in tourism and agriculture, along with the upgrade of local infrastructures.

The main objectives of the Programme are: Increasing the level of competitiveness in tourism, agriculture, labour market and education; improving infrastructures and transport networks to facilitate cross-border trade; improving cooperation in healthcare and social affairs; implementing cross-border projects on renewable energy and the management of natural resources.

2.2.1. Priority Axes of the Programme

The focal elements of the Programme have been summed up into three Priority Axes.

Priority axis 1: Competitiveness, knowledge and economic co-operation. This priority highlights the need of increasing competitiveness through cross-border cooperation, in order to modernize the labour market and to create more job opportunities.

Slovenian regions, in particular rural regions like Pomursko, still suffer from structural unemployment and long-term unemployment, especially youth and female unemployment. In 2004, in all Slovenian regions, except Pomursko, the female unemployment rate was over 50%. In addition, in Slovenia, the percentage of unemployed semi-skilled and unskilled workers was quite high.

---

5 European Commission, *European Territorial Cooperation 2007-2013: Cross-Border Cooperation between Austria and Slovenia – Operational Programme*, *op. cit.*, p. 12. Source: Östat (http://www.statistik.at) [accessed on 10th May 2013]; Hauptverband der österreichischen Sozialversicherung (http://www.hauptverband.at) [accessed on 12th May 2013]; Österreichische Sozialversicherung, http://www.sozialversicherung.at [accessed on 12th May 2013]; Arbeitsmarktservice Österreich, http://www.ams.at [accessed on 12th May 2013]; Statistical Yearbook of Republic of Slovenia, various years, http://www.stat.si/eng/pub_letopis_prva.asp [accessed on 2nd July 2013].

Another problem to tackle is the pay gap. Incomes in Austrian regions are higher than in Slovenia and this is often an incentive for Slovenians to work in Austrian enterprises located near the Slovenian border[6]. The current economic and financial crisis has worsened the unemployment situation in Slovenia.

This axis should allow small and medium enterprises (SMEs) to reach a higher level of development and dynamism, as the sphere of action of the majority of SMEs is still restricted to local territory. SMEs are specialized in sectors such as: mechanical/automotive industry, electronics, micro-electronics, wood processing, paper and plastics. Growing sectors are the food industry, software and eco-technologies. The Programme has supported cooperation between enterprises and research centres concerning new technologies (nanotech, IT, medicine, biomedicine, software) and new materials, in fields like wood processing, the automotive industry and environmental technologies. A certain number of SMEs have been able to create performing cross-border industrial clusters and networks, but cross-border cooperation in industrial clusters and networks is still limited and incomplete. The number of big enterprises is limited, even though multinational and big industrial enterprises can control a consistent share of industrial production in the regions. However, multinational enterprises may find a few difficulties in cooperating with local SMEs, as a large number of SMEs are still characterized by a low rate of internationalization; a low level of integration in transnational production chains and network; low export rates; and a limited geographical area of action[7]. Growth opportunities might also be obtained in rural regions, as both agriculture and tourism may be merged into one potential investment sector, in line with the core target of the first axis. In fact, tourism is considered one of the primary resources of the Austro-Slovenian cross-border area[8]. In addition, the promotion of cultural packages (museums, theatre, music, wine and culinary tourism), especially in the four main cities (Graz, Klagenfurt/Villach, Maribor and Ljubljana), is another important economic resource.

Priority axis 2: Sustainable and balanced development.

The key points of this axis are the safeguarding of the ecosystem and the use of renewable energy sources such as solar, biomass, water and

6 European Commission, *European Territorial Cooperation 2007-2013: Cross-Border Cooperation between Austria and Slovenia – Operational Programme*, *op. cit.*, pp. 20-23.

7 For further details about Slovenian SMEs, see also Gospodarske Zbornice Slovenije (GZS), Chamber of Commerce and Industry of Slovenia, http://eng.gzs.si/slo [accessed on 27th June 2013].

8 European Commission, *European Territorial Cooperation 2007-2013: Cross-Border Cooperation between Austria and Slovenia – Operational Programme*, *op. cit.*, p. 43.

wind, especially in Slovenia where the demand for energy has risen over the last twenty years. Austria has quite a long tradition in investing in new energy sources, such as hydro power and photovoltaic power. This priority axis also includes urban and regional development. This part of the Programme has encouraged the improvement of cross-border connections and new infrastructures to link the major cities of Austria and Slovenia, to achieve better access to cross-border opportunities for employees, students, customers and tourists. Concerning Austria, all major cities are linked by motorway networks, and secondary road networks are generally efficient. Slovenia has invested a lot in motorways, especially since 2004. Slovenian road infrastructures connecting the major cities have been enhanced considerably over the last ten years. However Slovenian railroad networks have not been improved and modernized adequately[9]. Investment is required in order to stimulate the improvement of Austro-Slovenian railroad networks and to facilitate connections between the major Austrian and Slovenian cities, especially on the Slovenian side.

Priority axis 3 is relevant to technical assistance.

2.3. The Forthcoming Cooperation Operational Programme (2014-2020)

By the end of 2013, the Operational Programme Slovenia-Austria for 2007-2013 will be over. Therefore, Slovenia and Austria have started to analyse the main issues for the new programming period 2014-2020, through regular meetings. During the "Development Conference" in Radeci[10], in Slovenia (12 March 2013), a list of priorities were defined: Enhancing the competitiveness of SMEs; protecting the environment (low-carbon economy, sustainable transport, bottlenecks elimination); investing in education (lifelong learning); supporting labour mobility and social inclusion. The Operational Programme 2014-2020 is supposed to make use of and to improve the priorities of the previous programme, and to coordinate strategic issues for further cross-border cooperation opportunities.

2.4. Euregio Project (Steiermark and North – Eastern Slovenia)

The Euregio Project for Steiermark and North – Eastern Slovenia (Evroregija in Slovenian) has been implemented in order to promote

[9] There are two main cross-border railway connections between the Austrian and Slovenian regions: Villach/Rosenbach – Jesenice and Spielfeld – Šentilj.

[10] The *Development Conference* was organized in Radenci (Slovenia) by Regional Development Agency Mura Ltd. on 12th March 2013. The conference was attended by 182 participants, as members of institutions, organizations and enterprises. Concerning the Operational Programme 2014-2020, see also the internet site of "Slovenia-Austria: European Territorial Cooperation": http://www.si-at.eu [accessed on 15th July 2013].

cross-border cooperation and trans-regional socio-economic development between Slovenia and Austria since the mid-1990s. Euregio is an instrument for bilateral cooperation (Austria – Slovenia) and for the coordination and implementation of pioneering cross-border projects concerning specific issues, such as: tourism, cultural packages, education and training, environmental protection, or even health care and assistance. It has been put into action through the Euregio Steiermark Coordination Office, Austrian Regional Management (Graz, Oststeiermark, West- and Südsteiermark, Voitsberg) and Slovenian agencies, such as: EIM (Ekonomski Institut Maribor); MRA (Mariborska Razvojna Agencija – Maribor Development Agency); RRA (Regionalna Razvojna Agencija – Regional Development Agency) Celje; RRA Koroška; RRA Mura; and SRC (Scientific Research Centre) Ptuj. The project has mainly focused on the creation of a cross-border network between municipalities, enterprises, institutions, non-government organizations and the local population, in order to develop innovative cross-border initiatives[11].

3. Cross-Border Cooperation between Italy and Slovenia: the Operational Programme 2007-2013

3.1. Geomorphology of the Territory Involved in the Programme

The Cross-border Operational Programme between Italy and Slovenia for the period 2007-2013 was approved by the European Commission in December 2007[12]. The 2007-2013 Operational Programme involves Italian and Slovenian areas that lie along their common border: The Italian provinces of Trieste, Gorizia, Udine, Venice, Rovigo, Padua, Ferrara and Ravenna; on the Slovenian side: Goriška, Obalno-Kraška and Gorenjska. The areas of Osrednjeslovenska in Slovenia, and the provinces of Pordenone and Treviso in Italy have taken part as an adjacent area[13].

The Operational Programme, which started in 2007, has focused on different types of landscapes, such as: Northern coastal regions along the Adriatic Sea; Alpine regions (Julian and Carnic Alps); the Carso/Kras calcareous rock region; valleys in the Emilia-Romagna and Veneto Regions; the Po river delta. In comparison to the previous programme (2000-2006), the 2007-2013 Programme area is much wider. The

---

11 Further details concerning Euregio/Evroregija Project are available at: http://www.evroregija.net (only in German and Slovenian) [accessed on 30th June 2013].

12 European Commission, *European Territorial Cooperation 2007-2013: Cross-Border Cooperation between Italy and Slovenia – Operational Programme*, CCI number: 2007CB163PO036.

13 These areas receive a maximum 20% of the ERDF funding allocated to the Programme.

territory involved in the 2007-2013 cooperation programme is 30,740 square kilometres, i.e. 62.9% larger than the 2000-2006 Programme area. Important rivers are Adige, Piave, Po, Tagliamento, Isonzo/Soča, Timavo/Reka[14] and Dragonia. All these rivers flow into the Adriatic Sea. The 30-km. river Dragonia has the delicate function of delimiting the Slovenian-Croatian border in Istria. In Italy the river Po has a leading role for the growth of the Italian fluvial transport.

Most of the territory of the Programme is subject to hydro geological risk, seismic phenomena and coastal erosion. Pollution levels as well as water contamination risks are elevated in the industrial areas of Porto Marghera (Venice), in the petro-chemical sites of Ravenna and Pontelagoscuro (Ferrara), and in the industrial port of Trieste.

3.2. Gorizia and Trieste: Two Border Cities

The 2007-2013 Programme area is also characterised by the presence of historically contentious border line cities: Gorizia and Trieste.

Gorizia was divided into two parts immediately after the end of World War II. In 1947, most of the urban area of Gorizia continued to be part of the Italian territory, while the Transalpina train station[15] and the Eastern peripheral districts of Gorizia (Solcano/Solkan, Prestava/Pristava, Valdirose/Rožna Dolina, Moncorona/Kromberk, Gorizia San Pietro/Šempeter pri Gorizi and Vrtojba), which were later identified as Nova Gorica, became part of the Yugoslavian territory. The demographic growth and the economic development of Nova Gorica were significant during the 1950s and the 1960s, as Tito invested various resources in the Nova Gorica project in order to challenge the Western capitalist economy. Since 1948, Nova Gorica was built up with the purpose of establishing a modern and efficient socialist city – a sort of ideal socialist city defying capitalist and commercial Gorizia. The Gorizia and Nova Gorica city centres are separated by nearly one kilometer. During the Cold War, the urban division symbol was the Transalpina Square: half of the square was in Italy (Gorizia), the other half, including the Transalpina train station, in Yugoslavia (Nova Gorica). This square was divided by a cement wall.

---

[14] The river Timavo/Reka is a 50-km. Karst river which flows nearly 40 kilometres underground.

[15] During the Austro-Hungarian Empire, Gorizia was directly linked to Vienna through the Transalpina train station. In 1947, this station and its railroad network became part of the Yugoslavian territory. Therefore the Transalpina station became officially the Nova Gorica station. At the moment Gorizia and Nova Gorica have their own stations and their own independent railroad networks. There is no railroad connection between the two cities. In terms of public transport, the two cities are linked exclusively by bus through Via San Gabriele (Gorizia) – Erjavceva Ulica (Nova Gorica).

However, the situation was not as dramatic as in Berlin. Since 2004, especially since Slovenia joined the Schengen area (21 December 2007), this cross-border square has been the freedom of movement emblem and one of the icons of the European integration process[16].

Concerning Trieste, immediately after the end of the Second World War, the territory of Trieste and North-Western Istria were split into two zones, along the Morgan Line:

- Zone A. This territory was under the control of Anglo-American forces. Zone A corresponded to the current Province of Trieste.
- Zone B. This included Capodistria/Koper, Pirano/Piran, Isola d'Istria/Izola, Umago/Umag and Cittanova/Novigrad. This zone was under Yugoslav People's Army control.

From the end of World War II until 1954, Trieste – the Free Territory of Trieste – was an independent state under UN protection. Trieste was definitively annexed to Italy in October 1954. The distance between the two coastal cities of Trieste and of Koper/Capodistria is marginal (almost 14 kilometres). Nowadays these two cities still belong to two different countries, but they are both part of the EU, so the minimal geographical distance between them should facilitate effective cross-border cooperation and help remove social, cultural and historical barriers. Besides, Trieste and Koper are two of the main harbours of the Adriatic Sea, so their proximity can also be a key factor in implementing common efficient maritime traffic strategies[17].

3.3. Economic Scenario

Concerning the economic context, in Italy, Emilia-Romagna, Veneto, Friuli-Venezia-Giulia and in Slovenia, Osrednjeslovensko, industry and service sectors are well developed. Agriculture is an important resource, as over 50% of the programme area is linked to the primary sector.

In Slovenia, in the region of Gorenjska more than 80% of the territory is cultivated, while in Italy agriculture is predominant in the provinces of Ferrara, Ravenna and Rovigo. Moving northward (i.e. Friuli), the percentage of territory cultivated tends to be inferior due to the

[16] Concerning an analysis of a long-term socio-economic reintegration of the two border areas (Gorizia – Nova Gorica), see M. Bufon, "Social Integration in the Italo-Slovene Gorizia Transborder Region", in *Journal of Economic and Social Geography (Tijdschrift voor economische en sociale geografie)*, Vol. 87, No. 3, June 1996, pp. 247-258.

[17] Concerning details on the twentieth century history of Trieste, see G. Sluga, *The Problem of Trieste and the Italo-Yugoslav Border*, Albany, State University of New York Press, 2001.

geo-morphological conformation of the territory (Julian and Carnic Alps, and Karst region)[18].

Tourism is another fundamental resource which has to be taken into consideration due to archaeological and religious architectural heritage existing on the territory (Ferrara, Padua, Ravenna, Udine and Venice).

Concerning education, in the Slovenian area of the programme, almost 70% of the population have reached a high level of education (diploma or degree), in comparison to 33% of the Italian population living in the programme regions. The percentage of graduates in the Slovenian area is nearly 15%, while on the Italian side the percentage is much lower (approximately 7%). In the Italian area, only 26% of the population between 20 and 24 years old has obtained a middle-school diploma[19].

Regarding linguistic minorities, in Italy, Slovenians live along the border with Slovenia, especially in the province of Trieste and Gorizia, while in Slovenia, Italians live in the coastal area. Both Italy and Slovenia have ratified the Council of Europe Framework Convention for the Protection of National Minorities. In Slovenia Italian minorities are protected by the Constitution[20].

3.4. Priority Axes

The total budget settled for the operational programme is 136 million euro. The European Regional Development Fund (ERDF) amounts to 116.2 million euro, which represents approximately 1.3% of the total EU

---

18 Karst region is also known as Karst Plateau. This area is a limestone plateau (Carso in Italian and Kras in Slovene).

19 Source of statistical figures: Statistični urad Republike Slovenije (Statistical Office of the Republic of Slovenia), http://www.stat.si [accessed on 2nd July 2013].

20 Article 64 of the Constitution of the Republic of Slovenia: Special Rights of the Autochthonous Italian and Hungarian National Communities in Slovenia. "The autochthonous Italian and Hungarian national communities and their members shall be guaranteed the right to use their national symbols freely and, in order to preserve their national identity, the right to establish organisations and develop economic, cultural, scientific and research activities, as well as activities in the field of public media and publishing. In accordance with laws, these two national communities and their members have the right to education and schooling in their own languages, as well as the right to establish and develop such education and schooling. The geographic areas in which bilingual schools are compulsory shall be established by law. These national communities and their members shall be guaranteed the right to foster relations with their nations of origin and their respective countries. The state shall provide material and moral support for the exercise of these rights". For details, see Slovenian Ministry of Interior, http://www.mnz.gov.si/en/minorities/italian_national_community [accessed on 3rd July 2013]. The Slovenian Constitution can be downloaded at: http://unpan1.un.org/intradoc/groups/public/documents/UNTC/UNPAN014895.pdf [accessed on 2nd July 2013].

investment earmarked for the European Territorial Cooperation Objective under the Cohesion Policy 2007-2013. The Operational Programme consists of four main priority axes.

Priority 1: Environment, Transport and Sustainable Territorial Integration. This priority is based on the increased use of renewable energy sources, the protection of bio-diversity, and the improvement of integrated transport systems, in accordance with the objectives of the Gothenburg Strategy of 2001 for sustainable development.

The main objective is the enhancement of transport networks and infrastructures, in order to ensure a superior level of integration between urban and rural areas. Besides, bordering areas must be joined with the rest of the inner territories, in order to avoid any type of isolation.

Priority 2: Competitiveness and a Knowledge-based Society. This priority is based on joint research and innovation initiatives in order to ensure environment protection as well as the development of a knowledge-based society in the programme region. The principal purposes can be summed up as follows: Increasing the competitiveness of enterprises (mainly SMEs); developing tourism (e.g. promotion of local/typical products from agriculture and creation of new holiday resorts); supporting the knowledge based economy. The cooperation between universities, enterprises and research institutes is central and indispensable in order to ensure technological development of Italian-Slovenian cross-border regions and to increase competitiveness. Knowledge transfer, innovation and scientific research are basic elements to create technological parks, enterprise incubators, and to set up hi-tech cross-border clusters.

Priority 3: Social Integration. One of the fundamental objectives of the Programme is to remove historical and cultural barriers, thus the Programme aims to improve communication and social and cultural cooperation between Italian and Slovenian regions, basically through the valorisation of cultural and historical heritage, and the respect of linguistic minorities.

Priority 4 is relevant to technical assistance.

4. Improvement of Transport Infrastructure and Sustainable Growth Opportunities in Slovenia

4.1. Operational Programme: "Development of Environment and Transport Infrastructure"

Since the so called "big-bang" enlargement in May 2004, the European Union has acquired a strategic role in Central and Eastern European countries. Slovenia, due to its central position inside Europe, is the heart of

Europe in terms of infrastructure networks and corridors, such as the Trans European Network (TEN): Corridor number 5, from Lisbon to Kiev; and Corridor number 10, from Northern Germany to Thessaloniki in Greece.

In 2007, the European Commission approved the Operational Programme "Development of environment and transport infrastructure" for Slovenia for the period 2007-2013 through Decision C (2007) 4081-1[21]. The core target of this programme is the enhancement of the main infrastructures and the development of sustainable mobility, in order to reduce carbon dioxide emissions. The Programme highlights the combination of infrastructure improvement and environmental protection as an essential measure to obtain sustainable growth and balanced regional development. The Programme is financed by the Cohesion Fund and the European Regional Development Fund (ERDF) under the Convergence objective.

The Programme takes into consideration the modernisation of 428 km of railways, especially along the TEN axis Lyons – Trieste – Ljubljana – Budapest – Ukrainian border. More to the point, the Ljubljana – Zidani Most – Maribor track is a key point in order to connect Slovenia with the rest of Europe and to allow faster connections inside the Italy – Austria – Slovenia triangle.

The double track section Ljubljana – Zidani Most – Maribor (155 kilometres) is part of the old "Southern Railway" line, which was built between 1846 and 1849 to reach Vienna from Trieste. Now this railway section is part of both Corridor number 5 and of Corridor number 10. The Ljubljana – Pragersko section (part of Corridor number 5) is essential to link Venice to Budapest. The Ljubljana – Zidani Most segment (part of Corridor number 10) is crucial to connect Salzburg (Austria) to Zagreb (Croatia), Belgrade (Serbia) and Sofia (Bulgaria), via Villach and Jesenice. The Zidani Most – Maribor section is necessary to link North-eastern Slovenia to the city of Graz in Austria, and then to proceed up to Vienna. In addition, the railway modernization process has included the purchase of new modern trains like the "Pendolino" (able to tilt on curves), new control equipment and a harmonization of regulatory norms. The purpose is to increase the average speed of trains and safety level.

4.2. Axes of the "Development of Environment and Transport Infrastructure" Programme

The "Development of environment and transport infrastructure" Programme is based on seven axes.

---

21 European Commission, *Operational Programme "Development of Environment and Transport Infrastructure"*, CCI number: 2007SI161PO002.

The first three axes focus on transport and the Trans-European Network strategy; the other four axes stress environment and sustainable growth, and technical assistance.

Priority axis 1: Railway infrastructure. This is financed through the Cohesion Fund. This priority axis deals with railway infrastructure modernisation in order to promote railway transport and to protect the environment.

Priority axis 2: Road and maritime infrastructure. This priority, which is financed through the Cohesion Fund, aims to reduce bottlenecks in order to avoid traffic congestion on the main roads. It also supports the enhancement of multimodal transport systems, the construction of intermodal transport terminals and the development of maritime routes (Motorways of the Sea).

Priority axis 3: Transport infrastructure. This priority axis assists the less developed and peripheral Slovenian regions. The goal is the improvement and the modernization of road and railway infrastructures in these areas in order to avoid isolation.

Priority axes 4 and 5 are about municipal waste management and environmental protection. (i.e. waste collection and separation, and urban waste water treatment).

Priority axis 6: Sustainable use of energy. This priority axis promotes the implementation of sustainable and efficient use of energy and the production of energy from renewable energy sources, in order to reduce greenhouse gas emissions, as required by the Kyoto Protocol.

Priority axis 7: Technical assistance.

5. Conclusion

The 2007-2013 and 2014-2020 Operational Programmes may be a turning point for Slovenian economic growth. These cooperation programmes are essential to overcoming regional development differences between Slovenian regions such as Podravsko, Pomursko and Koroško and the South-eastern Austrian regions or North-eastern Italian regions. However, there are several weak points that must be challenged.

First of all, concerning the territory, in order to obtain territorial sustainable integration, we have to bear in mind that the level of water consumption in some regions is extremely elevated. In addition, because of the soil morphology, some areas are at risk of earthquakes, floods and landslides, meaning that the ground must be monitored constantly.

Secondly, another problem to tackle is the lack of road infrastructure cohesion bringing about periodical traffic congestion in Ljubljana, Maribor and along the Slovenian coast. This deficiency comes out especially during the summer time around the main holiday resorts, i.e. the Italian-Slovenian border area (Trieste and Koper) and in the Istrian peninsula to access Croatia (Portoroz, Izola and Piran).

Thirdly, the low rate of competitiveness of the majority of Slovenian SMEs is another important problem that must be solved within the cooperation strategy. Several SMEs have shown an incapacity to face the internationalization of markets and they have only partially developed the ability to export products proficiently. In fact, most of the cross-border cooperation area enterprises are local dimension SMEs[22]. The number of high added value SMEs in the cross-border regions is still excessively low, due to the lack of research and investments in high technology sectors. Only a small percentage of the SMEs of the regions involved in the programme can be considered capital intensive firms.

Fourthly, the labour market needs some reform, especially in Slovenia, where the level of flexibility is still insufficient to compete in the global market. Nevertheless, Slovenian workers, thanks to the Schengen agreement, are able to move around the cross-border areas, especially commuting to Friuli Venezia Giulia or Austria, since, at the present time, the average income in Italy and Austria is generally higher than in Slovenia.

Last but not least, pertaining to social integration, an unsolved and unanswered problem is the presence of cultural and linguistic barriers. The Italian and Slovenian communities are still showing difficulties in cooperating efficiently because of historical and cultural obstacles.

---

22 A large number of SMEs in Slovenia consist of one or two employees, consequently the concept of micro-firms instead of small-medium enterprises could also be used.

Labour Market Migration in the German-Polish and in the Danish-German Border Region

Gerhard BESIER
Sigmund Neumann Institute, University of Dresden, Germany

Katarzyna STOKŁOSA
Department of Border Region Studies,
University of Southern Denmark

1. Polish Labour Market Migration into Germany

The political upheavals of 1989 unleashed a new wave of westwards migration from Poland. Far from a new phenomenon, Poland had already lost a considerable number of its inhabitants in the 1980s, who had sought to flee to London, Paris and the USA and thereby escape the Communist regime of their homeland. A further destination was West Germany, which received almost one million Poles[1]. Whilst the majority of emigrants moved for political reasons, economic considerations gradually emerged as the dominant cause of emigration during the second half of the 1980s. The end of Communism in the Eastern bloc meant that after 1989, labour migration accounted for the overwhelming majority of Polish emigration. Again, this was no new phenomenon. The German Democratic Republic and Poland had moved as early as the 1960s to sign an agreement facilitating the employment of Poles in the labour-hungry industries of the GDR. Such labour co-operation was made possible by the level of integration effected by COMECON. Seeking to distance itself from West Germany, the GDR was keen to drive on the integration of the COMECON member-states[2].

---

1 B. Kerski, "Hybride Identitäten. Migrationen aus Polen – Geschichte und Gegenwart", in Deutsches Polen-Institut (eds.), *Jahrbuch Polen 2010. Migration*, Wiesbaden, Harrassowitz Verlag, 2010, p. 12.

2 S. Röttger, "Ideologie, Integration und nationale Interessen – Argumentationsspielräume der DDR und Polens zwischen Westpolitik und Ostintegration in den 1970er Jahren", in D. Bingen, P. O. Loew, and N. Wolf (eds.), *Interesse und Konflikt. Zur politischen Ökonomie der deutsch-polnischen Beziehungen, 1900-2007*, Wiesbaden, Harrassowitz Verlag, 2008, pp. 284-285.

The majority of Polish workers employed in GDR factories were women living in close proximity to the German-Polish border. In this way, a large number of female Poles developed into commuters, living on the Polish side of the border and working in the GDR.

Good working relationships between Polish workers and their East German colleagues soon developed into friendships and a number of marriages. Nevertheless, prejudice remained and still exists in unified Germany. Following Poland's accession to both the EU in 2004 and the Schengen area in 2007, it is not uncommon for Poles to be disparaged in the German media as "illegal immigrants, people-smugglers and car thieves"[3].

The first part of this chapter seeks to compare and analyse cross-border Polish labour migration between Poland and the GDR before and after 1989. Concentrating less on establishing the facts, we will examine the prejudices and stereotypes which appear in connection with this phenomenon.

1.1. Polish Women in GDR Factories

The dynamic growth experienced by the GDR economy in the 1960s resulted in an ever-increasing labour hunger. This was accompanied by increasing levels of unemployment in the Polish border area, especially amongst women. Taking advantage of this situation, the GDR and Poland moved in 1966 to conclude an agreement permitting the employment of Polish workers, especially women, in GDR factories. The agreement guaranteed specific working conditions for Polish workers involving equal pay, bonuses and a range of other advantages. Polish labour migrants were also to be permitted to enjoy Polish public holidays. For their part, the Poles were required to learn German[4].

Polish women working in factories in the GDR played an important role in the appearance of Zgorzelec[5]. The majority of Polish labour commuters travelled on buses laid on for the purpose by their employer. One particular privilege accorded to Polish workers was their right to go shopping in the GDR, famed in the Eastern bloc for its supply

---

3 J. Judkowiak, "Die Rezeption der Schengen-Osterweiterung in der deutschen und polnischen Öffentlichkeit und ihre interkulturellen Aspekte", in H. Zschiedrich (ed.), *Wirtschaftliche Zusammenarbeit in Grenzregionen. Erwartungen – Bedingungen – Erfahrungen*, Berlin, Berliner Wissenschafts-Verlag, 2011, p. 403.

4 R. Röhr, "Die Beschäftigung polnischer Arbeitskräfte in der DDR 1966-1990; die vertraglichen Grundlagen und ihre Umsetzung", in *Archiv für Sozialgeschichte*, Vol. 42, 2002, pp. 211-236.

5 C. Anders, "Polnische Arbeitskräfte in Elektroschaltgerätewerk Görlitz", in U. Marquardt, and N. Faust (eds.), *Görlitz. Von der mittelalterlichen Handelsstadt zur Grenzstadt an der Neiße. Nahe ferne Nähe*, Görlitz and Zittau, Gunter Oettel, 2000, pp. 227-239.

situation and an opportunity denied to their neighbours working in Poland. For their part, Polish women brought flowers and vegetables, which enjoyed a high scarcity value in the GDR. Polish workers felt very comfortable in Germany, and their diaries and memories are full of references to good mutual relations between Polish and German women. When questioned today, those Polish women who worked in the GDR assure they always felt that they enjoyed the respect of their colleagues.

Polish women employed in GDR factories had a low level of education. This female preponderance in factories also influenced the character of the co-operation between Poland and the GDR. Zofia (born in 1947) spent the majority of her working life employed in a factory close to Görlitz, until the unification of Germany:

> I can't complain. Whatever happened I can't say a bad word about the Germans. If you asked them [her German colleagues] to buy something for us when we had a night shift or other or indeed at any other time, they always obliged. Sometimes they asked us to bring flowers for them – carnations from Poland. Because they didn't have any flowers. Sometimes we brought them tomatoes or cucumbers. We soon became good friends[6].

When asked about the German attitude towards Polish workers, she replied:

> Germans treated us very well. I can't complain. German women always organised something for the celebrations such as the New Year's party, and asked us what we wanted. They usually bought vodka and we prepared some snacks. You know how it is: Women's Day or somebody's birthday. I have very good memories! It was just like being in Poland[7].

Zofia was enthusiastic about her work. She also had a good opinion of the factory director and its board, as Polish workers were granted leave to celebrate Polish public holidays and were provided Polish-speaking medical care. Poles were also included in the system of bonus payments. Zofia remembered clearly that German bakeries opened at 6 a.m. – she bought delicious bread and cakes after a night-shift or just before starting the day-shift. Zofia mentioned several times that if such an opportunity existed today, she would not hesitate to take up the job she performed for thirty years.

The employment of Polish women also features in some of the interviews with East German women (especially of older generations) employed in factories in the border area in the same period. One female

6 Interview with Sofia P., 2010.

7 Interview with Sofia P., 2010.

inhabitant of Görlitz, born in 1932 told us: "We had a lot of Polish women working in Görlitz. We had a lot of factories"[8].

During the course of the interview, she emphasised that the majority of experiences with her Polish colleagues had been good. Rudolf (born in 1933) stressed the number of contacts made with the Polish women working in his factory. He remembered how Polish women crossed the border on a daily basis and that a number of friendships were made between Polish and German working women[9]. Mr. L. (born in 1937) also remembers the joint celebrations between German and Polish women such as those for Women's Day[10]. Werner (born in 1953), on the other hand, had predominantly negative memories of the period in which Polish workers were employed in the GDR factories. He mentioned fights in a dance hall in Görlitz, which as he believes, were provoked by Polish guest workers[11].

2. Polish Migration to Germany after 1990

The most important flow of Polish migration to Germany in the 1990s and 2000s involved seasonal workers. This trend was not affected by the Eastwards expansion of the EU, although the levels of migration to Great Britain and Ireland increased[12]. Despite lacking precise statistical data regarding Polish labour migration to Germany[13], it was calculated that over 300,000 Poles leave for Germany every year to take up paid employment[14]. Also of significant interest is the reverse development, with Poland becoming ever-more attractive for Germans seeking work[15]. After the USA and Switzerland, Poland ranks as the third-most popular destination for German economic migrants, even ahead of Austria. This

---

8 Interview with Frau G., 2008.

9 Interview with Rudolf B., 2008.

10 Interview with Mr. L., 2008.

11 Interview with Werner R., 2008.

12 P. Kaczmarczyk, "Arbeitskraftwanderung aus Polen – Die Erwartungen vor und die Realität nach der EU-Osterweiterung", in M. Nowicka (ed.), *Von Polen nach Deutschland und zurück. Die Arbeitsmigration und ihre Herausforderungen für Europa*, Bielefeld, transcript Verlag, 2007, pp. 100.

13 P. Kaczmarczyk, "Arbeitskraftwanderung aus Polen – Die Erwartungen vor und die Realität nach der EU-Osterweiterung", *op. cit.*, pp. 89-90.

14 E. Hönekopp, "Polnische Arbeitsmigranten auf dem Arbeitsmarkt in Deutschland zwei Jahre nach der Erweiterung", in M. Nowicka (ed.), *Von Polen nach Deutschland und zurück. Die Arbeitsmigration und ihre Herausforderungen für Europa*, Bielefeld, transcript Verlag, 2007, p. 47.

15 A. Jarosz, "Alltag im Grenzraum. Polen im Landkreis Uecker-Randow", in *Inter Finitimos. Jahrbuch zur deutsch-polnischen Beziehungsgeschichte*, Vol. 9, 2011, pp. 159-167.

development is to be viewed against relative Polish success in dealing with the international economic crisis[16].

The accession of Poland to the EU in May 2004 and more importantly, her entry into the Schengen area in December 2007 resulted in yet another increase in migration to Great Britain, Ireland and Germany. 2.5 million Poles left Poland at that time and started to look for a job abroad[17].

In Germany, this development resulted in growing fears amongst Germans that such levels of migration could endanger their jobs, mixed with concerns regarding the conditions under which low-paid Poles were forced to live. With low wages paid cash-in-hand and lacking any social security provision, many such immigrants were thought to live in abysmal conditions. Polish reports were dominated by the level of discrimination suffered by expatriate Poles and Polish firms on the markets of the old EU, especially in Germany[18].

Contemporary Polish labour migration in the German-Polish border region has spread fears amongst young Germans living in the border region of losing their jobs. Tom (born in 1987) is critical of the fact that Polish workers are prepared to work for less money, resulting in his opinion in the loss of jobs[19]. Silvana (born in 1982) accuses Polish workers of accepting low-paid work and thus taking the jobs from better-qualified Germans[20]. Older inhabitants of Görlitz, such as Walter (born 1949), are ashamed that Poles and other economic migrants from Eastern Europe are forced to do the same work for much lower wages[21].

Interesting appears the fact that after the opening of the labour market for migrants in Germany on the 1st of May 2011, migration into Germany was much lower than expected, although 1.5 million Poles live in other EU countries today[22].

16 W. Smoczyński, "Ile Niemca w Polaku [How German is a Pole]", in *Polityka*, Vol. 16, 2012, p. 22.

17 V. Trappmann, "Flexibilität zu welchem Preis? Arbeit in Polen", in Deutsches Polen-Institut (ed.), *Jahrbuch Polen 2013. Arbeitswelt*, Wiesbaden, Harrassowitz Verlag, 2013, p. 20.

18 M. Nowicka, "Introduction to: Migration als Herausforderung für Europa", in M. Nowicka (ed.), *Von Polen nach Deutschland und zurück. Die Arbeitsmigration und ihre Herausforderungen für Europa*, Bielefeld, transcript Verlag, 2007, p. 8.

19 Interview with Tom M., 2008.

20 Interview with Silvana B., 2008.

21 Interview with Mr. Walter, 2008.

22 M. Duszczyk, "Arbeitsmarkt und Arbeitsmarktpolitik in Polen", in Deutsches Polen-Institut (ed.), *Jahrbuch Polen 2013. Arbeitswelt*, Wiesbaden, Harrassowitz Verlag, pp. 16-17.

3. Prejudices in the German-Polish Border Region then and now

Anti-Polish prejudice existed during the Communist period and continues today. Whilst the memories of our Polish interview partners of their work in the GDR remain predominantly positive, some of our German interview partners expressed a certain level of scepticism towards Poles and all Polish things. A widely-held anti-Polish prejudice, common both in the GDR and post-unification Germany is that of the lazy, yet sly Polish worker.

A positive factor in shaping relationships (both then and today) was the factor of "friendship". Private contacts played (and indeed, continue to play) a very important role in the development of relationships, irrespective of the political system under which they function. Proving themselves to be of far greater importance than any economic data, the formation of friendships between Polish guest workers and Germans were and are able to combat prejudice and stereotypes.

4. Labour Market Migration in the Danish-German Border Region

This topic could be addressed from a number of perspectives. We could adopt a historical approach tracing the long history of German migration to Denmark, something which only became problematic following the rise of nationalism in the 19th century[23]. There is little to say from the other perspective, as migration from Denmark to Germany never occurred to any great extent. The current situation is not particularly characterised by any level of permanent migration in either direction, but by the phenomenon of cross-border commuting, i.e. regular cross-border travel for the purpose of taking up paid employment in one country, whilst remaining resident in the other[24]. Formal immigration is much rarer. From 2008 to 2011, between 18,000 and 22,000 Danish citizens lived in Germany (the figures for 2008: 18,001; 2009: 20,387; 2010: 21,114; 2011: 21,630)[25]; in the same period, roughly the same number of Germans (2008: 20,201; 2009: 20,444; 2010: 20,283; 2011: 20,428) lived in Denmark[26]. Comparison of

---

23 G. Snell, "Vom Miteinander zum Nebeneinander. Die Immigration von Deutschen nach Dänemark", in *Nordeuropaforum. Zeitschrift für Politik, Wirtschaft und Kultur*, Vol. 2, Berlin, 2000, pp. 3-15.

24 T. Buch *et al.*, "Grenzpendeln in der deutsch-dänischen Grenzregion. Entwicklung und Struktur 1998-2005", in *IAB regional. Berichte und Analysen (IAB Nord)*, Vol. 4, 2008.

25 Eurostat, http://appsso.eurostat.ec.europa.eu/nui/submitViewTableAction.do [accessed on 9th September 2012].

26 M. Wirsing, "Über eine halbe Million Deutsche leben in den Nachbarländern", in Statistisches Bundesamt (eds.), *Bevölkerung und Erwerbstätigkeit. Ausländische Bevölkerung, Ergebnisse des Ausländerzentralregisters*, Wiesbaden, 2012, https://

the population fluctuation under consideration of arrivals and departures reveals a high level of stability in terms of Danes in Germany (between 66 and 450 people) and a higher level for Germans in Denmark, with figures fluctuating between 250 and 2,500 people per year.

In 2005, 2,600 people left Germany for Southern Jutland. 2011 saw 4,000 people leaving Schleswig for Southern Jutland – the figures for Denmark as a whole amount to some 12,000 people. In the same period, only 540 people left Denmark for Schleswig in 2005; depending on the statistics other figures show 1,000 people.

The great majority of the cases of migration involved labour migration. The Scandinavian lands, "especially Norway and Denmark currently exhibit a very high demand for German workers, especially craftsmen and those engaged in metal, construction and ancillary construction trades […] not least due to the dual education system in which they completed their training"[27]. A full treatment of this topic requires a comparison of the migration and immigration policies of both countries, even if this largely involves immigration from non-Western source lands[28]. Finally, the social and economic significance of cross-border German-Danish shopping tourism also means that this issue cannot be ignored[29].

5. Cross-Border Commuting in the German-Danish Border Region

The integration of the regional employment market, i.e. the cross-border placement of labour plays an important part, numbers amongst the principal aspects of cross-border co-operation in recent years. A joint German-Danish study, which was published in 2008 and focused on the development and structure of border commuting, established that in 2005, some 2,500 people commuted from Germany to Southern Jutland, whilst only 540 people undertook the journey in the opposite direction. The figures for commuting Danes have remained largely stable, whilst

www.destatis.de/DE/PresseService/Presse/Pressemitteilungen/2009/12/PD09_494_124.html [Press release from 17th December 2009, accessed on 10th May 2012].

27 S. Mau *et al.*, "Die Arbeitsmigration von Deutschen in Europa – Erste Ergebnisse einer qualitativen Untersuchung", in K. S. Rehberg (ed.), *Die Natur der Gesellschaft – Verhandlungen des 33. Kongresses der Deutschen Soziologie in Kassel 2006*, CD-ROM, Frankfurt, Campus, 2006, p. 4476.

28 K. F. Zimmermann, and H. Hinte, *Zuwanderung und Arbeitsmarkt. Deutschland und Dänemark im Vergleich*, Heidelberg, Springer, 2005.

29 K. Wiegand, *Drei Jahrzehnte Reiseverkehr und Einkaufstourismus in der deutsch-dänischen Grenzregion – eine zusammenfassende Bilanz für die Zeit von 1957 bis 1987/88*, Flensburg, Forschungsstelle für Regionale Landeskunde, 1988; K. Wiegand, "Drei Jahrzehnte Einkaufstourismus über die deutsch-dänische Grenze", in *Geographische Rundschau*, Vol. 42, 1990, pp. 286-290 (summarised version).

the number of Germans working in Denmark has risen greatly since 2001. 2005 saw some 540 people commute from Denmark to Schleswig and some 2,600 from Germany to Southern Jutland[30]. Estimations place the untapped potential of cross-border commuters at some 14,000[31]. The relatively high level of unemployment present in Germany until 2009 acted as a "push factor" away from the German employment market; the effect of which was compounded by differences in wage levels, especially for poorly-qualified tasks. This represents the Danish "pull factor". Analysed according to gender, commuters on both sides of the border show a slight preponderance of males to females (60 to 40 per cent respectively). Moreover, those commuting from Germany to Denmark are on average younger than those commuting in the other direction. Not all border commuters (only between 60-70 per cent) hold the citizenship of the country in which they live: some originate from third countries. A higher proportion of border commuters from Germany number amongst the relatively lower income classes, whilst those employees travelling from Denmark to Germany earn a relatively high income. This points to the various branches from which commuters are drawn – those commuting from Denmark to Germany are concentrated in the processing industries, the wholesale and retail industries and service providing industries. Commuters from South to North are represented in all branches of industry. The significant growth in cross-border commuter integration results in a positive reduction of imbalances between labour markets. Such developments in the border area could indeed serve as a model for non-border regions; many regions in Europe still demonstrate a high level of labour inflexibility, expressed in an unwillingness to cross international borders to take up an attractive employment opportunity. Indeed, only 1.5 per cent of the total population of Europe has resolved to take this step. It is entirely possible that recent developments "push" and "pull" these effects more strongly from South to North. Nevertheless, it remains clear that cross-border commuting represents a specific form of labour mobility as such commuters are able to remain at home.

The reasons for this conventionally low level of labour mobility are usually to be found in the differences between the tax and social security systems, linguistic barriers and cultural differences. Reporting in 2006, a German-Danish working group seeking to promote cross-border labour mobility noted that administrative hurdles in the border region between

30 T. Buch *et al.*, "Grenzpendeln in der deutsch-dänischen Grenzregion, Entwicklung und Struktur 1998-2005", *op. cit.*, p. 12.

31 U. Barten, H. Hermann, and M. Klatt, *Barrierer og potentialer i det grænseoverskridendende samarbejde i Fmern Baelt Regionen*, Rapport af Institut for Grænsregionsforskning – Syddansk Universitet og Institut für Regionalforschung (Christian-Albrechts-Universität Kiel), 2006.

Schleswig and Southern Jutland also represent a significant handicap to cross-border mobility[32]. The situation is characterised by the lack of a long-term strategy to entrench and extend current patterns of labour market integration. In terms of measures designed to reduce linguistic and cultural barriers, it is clear that Southern Denmark is making a much more concerted effort in this direction than Schleswig.

6. A Comparison of Immigration Policies

Germany not only attracts a far greater number of labour migrants than Denmark, but is also more successful in her attempts to integrate the newly-arrived in her labour market[33]. A greater proportion of migrants to Denmark fail to end their dependence on the Danish social security system, a trend following from the conditions favouring low-income groups. The Danish system of unemployment benefits provides those on low incomes with a higher ratio of income replacement than in Germany, which acts as a disincentive to taking up paid employment. Nevertheless, the system of controls established in Denmark is far stricter than that established in Germany and exerts greater pressures to complete courses of further training.

The course taken by the history of migration in both Germany and Denmark has exhibited considerable similarities. After a phase of migrant recruitment, in the early 1970s political considerations influenced the decision taken to stop any further acquisition in favour not only of allowing the family members of the already-arrived, but also asylum seekers to enter the country[34]. In the meantime, both countries have implemented measures to effect a considerable restriction in levels of immigration on the part of low-qualified individuals. Starting in 2002, Denmark has introduced step by step legislation[35] which has resulted in Europe's strictest aliens and immigration law. Since 2009, the government provides all non-Western immigrants leaving Denmark unilaterally to return to their original homeland, with a one-off payment of €13,500. Further legislation was passed in November 2010 to place

[32] T. Buch *et al.*, "Grenzpendeln in der deutsch-dänischen Grenzregion, Entwicklung und Struktur 1998-2005", *op. cit.*, p. 55.

[33] K. F. Zimmermann, and H. Hinte, *Zuwanderung und Arbeitsmarkt. Deutschland und Dänemark im Vergleich*, *op. cit.*

[34] M. Berlinghoff, "Der europäisierte Anwerbestopp", in J. Oltmer, A. Kreinbrink, and C. Sans Díaz (eds.), *Das "Gastarbeiter"-System. Arbeitsmigration und ihre Folgen in der Bundesrepublik Deutschland und Westeuropa*, München, Oldenbourg Verlag, 2012, pp. 149-164.

[35] *Migration und Bevölkerung Newsletter* No. 6, 2002; *Migration und Bevölkerung Newsletter* No. 5, 2011; *Spiegel Online*, 28/04/2011.

further restrictions on subsequent immigration for the purpose of bringing married couples together.

Both countries have accorded considerable privileges to certain groups amongst the immigrants: ethnic German repatriates in Germany and Scandinavians in Denmark. The percentage of non-Western immigrants and refugees is higher in Denmark than in Germany. These groups occupy the lowest end of the labour market hierarchy in both countries. Measured against the total population, the percentage of foreign nationals living in Denmark (5 per cent) is only half that of the figure for Germany. Nevertheless, Danish statistics, counting those immigrants who have taken German citizenship, show a quota of almost 10 per cent, thus pulling level with the situation in Germany.

At the end of the 20th century, both Denmark and Germany tightened their immigration legislation in a highly similar fashion. In comparison to Germany, however, Denmark made greater efforts to effect the cultural integration of her immigrant population. Moving in 1999, she established a three-year system of integration and language courses with a labour market orientation. Denmark has also registered greater success than Germany in her efforts to establish a system of training and skill enhancement for immigrants. This greater level of success is especially apparent in the employment figures for "second generation migrants" who have demonstrated a greater ability than their German counterparts to emancipate themselves from the educational and social status of their parents. Despite such levels of success, there are significant differences of integrational achievement in terms of the original nationality of the immigrant population. In both countries, pronounced religiosity and the fact of socialization in a small town are two factors hindering the acquisition of "human capital" amongst the migrant population of both nations.

Traditional conceptions and gender models have a much greater and more restrictive effect on women in terms of education and training in Germany than they do in Denmark. Both countries currently exhibit a pronounced trend towards selective immigration combined with targeted measures of cultural integration based on labour-market requirements. The response of both states to a shortage in skilled workers has concentrated on the targeted recruitment of highly-qualified immigrants. The restriction of further immigration of low-qualified workers from non-Western countries would considerably unburden the social system, especially in Denmark.

At the beginning of 2011, the far-right populist Danish People's Party launched a campaign to restrict further low-qualified immigration from non-Western countries in Denmark. A ministerial report calculated that the current restrictions on immigration established in Denmark over the last ten years had resulted in savings amounting to € 6-7 billion[36]. Nevertheless, such voices have become much quieter since the voting out of the centre-right coalition (itself requiring the parliamentary toleration of the People's Party) in 2011. Many observers see that the continuing successes in their efforts to integrate migrants into Danish society will result in improvement in the peaceful and successful coexistence of "old" and "new" Danes[37].

7. Shopping Tourism

2010 also saw Germany maintain her position as the number one destination for Danish tourists[38]. The average duration of a Danish visit to Germany amounted to 3.7 nights; and the most popular type of holiday (taken by 47 per cent of tourists) was a town break[39]. In addition to the perception of Germany as "quick and easy to reach" (67 per cent), 43 per cent of all travellers praised Germany for her "good shopping facilities"[40]. In 2010, Danish tourists spent around 1 billion Euros in Germany. Such activities stand in a long tradition of shopping tourism over the German-Danish border, itself resulting from the changing price advantages enjoyed by a number of products[41]. Since the "butter year" of 1957 – when Danish butter was sold at prices one third lower than those in Germany – the levels of shopping tourism have increased in uneven peaks. The repeatedly-heralded economic harmonization set to follow from Danish accession to the European Community in 1973 did not manifest itself as expected. This ensured that shopping tourism continues to remain an important economic factor for this thinly-populated, under-industrialised border area, itself plagued by traditionally high levels of unemployment[42].

---

36 *Spiegel Online*, 05/2011.

37 H. K. Rasmussen, *Den danske stamme; en befolkningshistorie*, Copenhagen, Politikens Forlags, 2008; H. K. Rasmussen, *No Entry: Immigration Policy in Europe*, Copenhagen, Handelshojskolens Forlag, 1997; P. Nannestad, *Immigration as a Challenge to the Danish Welfare State?*, Paper prepared for presentation at Bar-Ilan University, Tel Aviv, Israel, 3rd November 2003.

38 DZT, 2011.

39 DZT, 2011.

40 DZT, 2011.

41 K. Wiegand, *Drei Jahrzehnte Reiseverkehr und Einkaufstouriusmus in der deutsch-dänischen Grenzregion – eine zusammenfassende Bilanz für die Zeit von 1957 bis 1987/88, op. cit.*

42 K. Wiegand, "Drei Jahrzehnte Einkaufstourismus über die deutsch-dänische Grenze", *op. cit.*

8. Conclusions

In each border region there are specific circumstances and problems. With regard to the German-Polish border region, we stressed the historical factor and that of cultural differences and prejudices. Those factors are present in the German-Danish border region as well, of course. But keeping in mind that living together with both sides in freedom and peace has lasted for approximately 70 years, we have decided to emphasise the economic factor in this case. Whereas a sense of "otherness" stimulates and vitalises the mutual cooperation in the German-Polish border region, putative familiarity between the German-Danish border people has got a certain tendency of mutual tiredness and indolence. This might be valid for the cultural as well as for the economic sphere.

Informal Cross-Border Cooperation in the Baltic Sea Area

Identity and Allegiance as a Hindrance and Possibility

Thomas LUNDÉN

Centre for Baltic and East European Studies, Södertörn University

1. Introduction

Informal cross-border cooperation, with a wide definition, includes all sorts of local actions involving actors from both sides of an international boundary, with the exception of those acts that are based on formal agreements backed by the two state governments or other international agencies. There is of course a wide intermediary zone, e.g. informal agreements between adjacent border municipalities. Included in the definition are, for example, cross-border marriages, football matches between the neighbouring border guards, common walking trails, use of summer houses on the other side and the use of the media "of the other side".

In all these actions – and the lack of them – questions of identity and allegiance in relation to the local borderland and to the respective territorial states are raised.

In an old standard dictionary of concepts, Roget's Thesaurus (1953/56)[1], *identity* is under the section Relation: Absolute relation. As concepts related to identity are mentioned:

Noun: *sameness, ditto, homogeneity, unity, coincidence, coalescence, convertibility, equality, self, oneself, identification, monotony, tautology, synonym, homonym, facsimile, same, self-, very-, one and the same, actual thing, no other*. Verbs, adjectives and adverbs reflect the same set of words. According to the Swedish National Encyclopedia there are at least three different meanings of the word identity: one is a total conformity between two propositions, e.g. a=a; another, for example,

[1] P. M. Roget, *Roget's Thesaurus*, London, Penguin, 1953 (56), p. 5.

is an ascertained existence of a person (e.g. through an identification card)[2].

2. Identity as Belonging

But obviously, when talking about the identity of an area, or rather its population, we are asking for the third meaning, a consciousness of belonging to a group, a territory or a culture. My identity is my "true belonging". Identity thus clearly reflects some relation. In our meaning of the word, identity is the fact that people refer themselves to a group of people or to an organisation to which they think, or say, they belong. This group of people could be a *nation*, whether it is thought of in terms of a state or a people. But it could also be a religious group, a football team, a class (in the Marxist sense of the word) or a network of individuals with the same interests, e.g. radio amateurs or Esperantists. *Ethnicity*, as discussed by Barth, is an ascription into a group of some perceived homogeneity, but most social scientists now see this ascription as a result of information and indoctrination rather than a primordial difference in relation to non-members[3]. Ethnic group often denotes a weaker self-image than a "people" (German *Volk*) or a "nation", in the cultural definition of this misused word. Nation here means a group of people who regard themselves as a nation that forms, or should form, a state (the *Kulturnation* concept) and not the French and US American definition where nation and state seem to converge. This difference, in turn, is related to identification being perceived as related to the territory with its population or to the population with its territory – *Jus solis*: everybody legally living in the state territory will be accepted as member of the nation/state, irrespective of origin or ethnicity (the "French model") and *jus sanguinis*, everybody with the right ethnic origins will be accepted into the nation/state, irrespective of their territorial origin (the "German model"). These two concepts have strongly influenced the way people define others, and also themselves. And in turn, these often conflicting concepts have influenced the geopolitical fate of the territorial states of Europe, particularly since the end of World War I[4].

One level of identification is thus nationalism, the feeling of identification with a specified group of people that forms a state, or

2 "Identitet", in *Nationalencyklopedin*, Nionde Bandet, Höganäs, Bra Böcker, 1992, p. 342.

3 F. Barth (ed.), *Ethnic Groups and Boundaries: The Social Organization of Culture Difference*, Oslo, Universitetsforlaget, 1969.

4 T. Lundén, "Language, Ethnicity and Boundary Changes in 20th Century Europe", in E. Medijainen, and O. Mertelsmann (eds.), *Border Changes in 20th Century Europe*, Tartu Studies in Contemporary History, Berlin, LIT Verlag, 2010, pp. 9-20.

intends to do it. I am using the word nationalism in a neutral sense, without any negative or positive connotations. Benedict Anderson[5] defined nationalism as the feeling of belonging to an *imagined community*. Most modern states try to make a nation out of its population. This, in turn, means that at a certain level people's identity should be with the territorial state they live in. Karl W. Deutsch, in his famous book *Nationalism and Social Communication* pointed at the fact that the state, or rather its government, tries to make its territory a region of communication, where internal links and nodes are stronger and more natural than those beyond the boundaries of the state[6]. If successful, the inhabitants of the state territory will feel allegiance to its territorial organisation irrespective of its ethnic or linguistic composition. The obvious case is Switzerland. This is, of course, a functionalist and constructivist view. In states with a weak basis for cohesion, strong efforts have to be made to reach a level where, for most people, domestic contacts seem more natural than those with e.g. kin-ethnic groups beyond the border. Brubaker calls such states *nationalizing states*[7]. We have here a third variant of the "law" of nation-building: making a nation out of a mixed population in a state territory often derived by coincidence or coercion.

For the rulers, there are simply speaking two ways of attaining individual identification with the state territory: by *inclusion* or *exclusion*: in the first case, the population, irrespective of ethnic or other origin, is either encouraged or coerced to feel allegiance. Take the example of Germany before 1933: if you were Danish, Polish or Jewish, you could switch to German language, German culture and Christian faith, and you would become accepted and welcomed as German[8]. During the reign of National Socialism: if you were not defined as member of the German *Volk*, you would meet exclusion, by repulsion or extermination. There were, of course, matters of definition in cases of cross-pressures: in the Danish borderland under Nazi German occupation, two Nazi parties were struggling for local hegemony: the party of the German minority, urging for territorial inclusion of Northern Schleswig into Germany, and the Danish party, trying to get acceptance as the party of an independent Germanic Nazi

---

5 B. Anderson, *Imagined Communities. Reflections on the Origin and Spread of Nationalism*, London and New York, Verso, 1991.

6 K. W. Deutsch, *Nationalism and Social Communication*, Cambridge (Mass.), Technology Press, 1953.

7 R. Brubaker, *Nationalism Reframed. Nationhood and the National Question in the New Europe*, Cambridge, Cambridge University Press, 1996.

8 T. Lundén, *On the Boundary. About Humans at the End of Territory*, Stockholm, Läromedel från Södertörns Högskola, 2004, pp. 50-51.

Denmark under the leadership of Germany. Ideology was formally the same, but "national" and consequently, territorial identity differed[9].

3. Identity in Border Communities

Borders, or *boundaries*, are visible or invisible lines drawn through a landscape, between people or in people's imagination. This chapter will concentrate on the physical boundaries separating territories and their effects on people's identities, in other words the relation between physical territories and people's concept of their belonging.

In border communities, people are *confronted*. One such confrontation is the issue of identity. Even if they feel a sense of belonging to the nationality (in itself a dubious word) of the "nation-state" (also a dubious word) they belong to, they will have another territory with other possible identities at close distance. If a majority of people in the middle of a large state feel a safe identity (at a certain level, i.e. in terms of nationality/ethnic group of the nation-state that this group represents), border dwellers will at least reflect over the spatial line that made them different from people on the other side of the line. They often develop a way of behaviour to express (what they think it is) their nationality, e.g. by using symbols – or by not using them, as will be discussed below.

In a border area, this identification can thus become questioned for two reasons. One is that the border delimitation cuts through an area populated by people of the same "nation of will". Part of the "nation" is beyond their "nation-state". Another aspect is that in most European states, boundaries are permeable, and feelings of loyalty may be vague or divided. In the case of Norway-Sweden, linguistic differences are small and while national awareness is taken for granted on both sides, local transmigrants (often seeking work) were not seen as "aliens" and were ethnically assimilated within a generation[10]. In the western part of the US-Canadian borderland, an area of common language, transmigrants into Canada were found to quickly adopt both Canadian citizenship and loyalty, but in this case the move was often caused by a "choice of political culture", especially in rejection of US military interventions[11].

---

9 S. Werther, *SS-Vision und Grenzland-Realität. Vom Umgang dänischer und "volks-deutscher" Nationalsozialisten in Sönderjylland mit der "grossgermanischen" Ideologie der SS*, Stockholm, Stockholm Studies in History 95, 2012.

10 T. Lundén, *On the Boundary. About Humans at the End of Territory*, *op. cit.*, pp. 179-180.

11 S. W. Hardwick, and G. Mansfield, "Discourse, Identity and "Homeland as Other" at the Borderlands", in *Annals of the Association of American Geographers*, Vol. 99, No. 2, 2009, pp. 383-405.

From the point of departure of local inhabitants, two principles of spatial relations are at work, the territorial and the functional one. In Hägerstrand's words:

> [E]very large group of human beings is subjected to a tension between two principles of integration, the *territorial* and the *functional* modes. In the first mode, *nearness* is the supreme category. Thinking, loyalty and acting are highly place-bound. In the functional mode of integration, *similarity* is the supreme category. Thinking, loyalty and action become of a 'non-place' kind and unite what is similar in function over wide geographical areas[12].

In relation to a state boundary, the nearness principle can be truncated by barrier effects, while in the functional mode similarities can be enclosed by state territorial definitions, as e.g. the state railway networks, or global without borders, as e.g. organisations of criminals or philatelists. Usually even a functional organisation has a state territorial level, related to the legal system of each state.

It can be added that the territorial principle, depending on nearness, is related to the market aspect, where distance is a cost, whereas the functional principle is more related to the political aspect, where membership and hierarchical ordering is more important than proximity. The dichotomy also resembles Tönnies' (1887) division into *Gemeinschaft* and *Gesellschaft*[13].

4. Contacts and Identity in Open Borderlands

There is a range of permeability from the hermetically closed boundary to the total openness, where the transgression is free along the boundary line. In the Norwegian-Swedish borderland, people can since long cross without passports and large areas are just forest lands with little habitation. Over history, there were many marriages crossing the ethnic divide (which formally clear-cut is in terms of communication very vague, at least with regard to language, as Swedish and Norwegian could almost be seen as dialects of Scandinavian rather than distinct languages). There was also a migration, mainly from Sweden into Norway. In these cases, incoming "ethnic Swedes" were assimilated into the majority population within two generations[14].

---

12 T. Hägerstrand, "Decentralization and Radio Broadcasting: On the 'Possibility Space' of a Communication Technology", in *European Journal of Communication*, Vol. 1, 1986, pp. 7-26.

13 F. Tönnies, *Gemeinschaft und Gesellschaft. Abhandlung des Communismus und des Sozialismus als empirische Kulturformen*, Leipzig, 1887.

14 T. Lundén, *On the Boundary. About Humans at the End of Territory*, *op. cit.*, pp. 174-185.

Many Norwegians have summer houses on the Swedish side. In both countries, the official flag of the state is used on flagpoles of summer houses. But in most cases, Norwegians with summer houses in Sweden would refrain from using the Norwegian flag on their plots. Flags also have a slightly different connotation. In Norway, the flag is associated with the national, i.e. state, independence, reflecting the fact that for many hundreds of years, Norway was under Danish rule, while later forced into a (rather weak) union with Sweden 1814-1905. The Swedish flag today is most associated with summer, vacation, birthday celebrations etc., as the Swedish statehood was never challenged within the present territory of the realm. In later years, however, the flag has been used ("kidnapped") by nationalist extremists in order to underline difference; which is in relation to immigrants and foreigners. In some working-class suburbs of greater cities, small Swedish flags at balconies and walls mark the presence of inhabitants of Swedish ethnicity, in difference to those of "the other".

5. Identity in Borderlands of Mixed but Defined Ethnicity

Borderlands are often areas of mixed ethnicity. People have ethnic or linguistic links to people on the other side. Many of Europe's recognised "national minorities" inhabit borderlands. Depending on their own culture and on the policies of their home state and their "kin-state", they may develop mixed or often split identities. In cases of a clear ethnic division, patterns of identity may take very clear forms.

An example of mixed but sharply defined ethnicity is from Denmark-Germany. A big *Dannebrog*, the national flag of Denmark, is flying at Skamlingsbanken, a hill located just where the state boundary between Germany and Denmark was drawn after the disastrous war of 1864-66, when Denmark lost its Duchy of Slesvig with Danish population. The flag is used by the Danes to signal togetherness, and the whole Danish borderland is full of such symbols. The German minority is conspicuous by their lack of flags: in the Danish-German borderlands, it is easy to see which house belongs to a Dane or a German. On the Danish side, houses with Danes have a flagpole, while the "German" houses stick out by not having one. If they had, they would have the choice of offending the majority by using the German flag, or denying their nationality by flying the *Dannebrog*. By the way, using the state flag on a private plot is not common in Germany, where the flag is a state cultural symbol[15].

In Estonia, the eastern border town of Narva is only divided from the neighbouring Ivangorod in Russia by a river. The inhabitants of Narva can

---

15 P. Wolf, and K. Svalastoga, *En by ved grænsen*, København, Gyldendals, 1963; T. Lundén, *On the Boundary. About Humans at the End of Territory*, *op. cit.*, p. 107.

be divided into three rather equal categories according to *citizenship*: some 40 per cent having acquired Estonian citizenship, 30 per cent are Russian citizens, while the remainder is considered aliens who require a special permit, which was until recently also valid for crossing the boundary. However, citizenship has very little to do with identity and allegiance. Almost every inhabitant of Narva and its domestic surroundings have Russian as a native language. In terms of identity, most Russians of Narva seem to identify with the Russian ethnic nation, but they see each other as Estonian Russians, different from *Rossiyanie*. Few of them seem to want to leave for Russia, young people usually want to "go west" and prefer learning English to learning Estonian. However, with increasing knowledge of the nation-state language, Estonian, they may identify with their country while at the same time they would not deny their ethnicity[16]. The Narva and Narva-Jõesuu area is full of symbols related to the local history of very different geopolitical regimes. These symbols were often erected to strengthen solidarity or identification with the reigning power. The inhabitant's reactions vary from enthusiastic endorsement to contempt or ignorance[17].

6. Areas of Mixed Identities

Minorities may also develop mixed or split identities. In several of the European "nation-states", border minorities have been oppressed linguistically. Their languages have been banned from schools and/or denounced in other ways. In the Swedish northern borderlands of the Torne River Valley, Finnish was spoken by the local population. However, after a long time of linguistic suppression, many people denied the value of the native language, and they did not try to speak it to their children. One reason for this was the belief that "you have to be Swedish, as we live in Sweden", and "Swedish people do not speak Finnish". After several generations of negligence, the original Finnish dialect has been weakened, mixed with Swedish and in the countryside many people say that their Finnish dialect is not Finnish but, *Meänkieli* (our language). This version was formally declared a language separate from Finnish. This type of denial of language, or of ethnicity that can also be found in German speaking Alsace (France) and in Slovene-speaking border areas

16 T. Lundén, and D. Zalamans, "'National Allegiance' and Spatial Behavior in Baltic Boundary Twin Towns", in *Journal of Baltic Studies* XXXIII, Vol. 2, 2002, pp. 177-198; T. Lundén, "Valga-Valka, Narva-Ivangorod. Estonia's Divided Border Cities-Cooperation and Conflict Within and Beyond the EU", in J. Janczak (ed.), *Conflict and Cooperation in Divided Towns and Cities*, Berlin, Logos Verlag, 2009, pp. 133-149.

17 S. Burch, and D. J. Smith, "Empty Spaces and the Value of Symbols: Estonia's 'War of Monuments' from another Angle", in *Europe-Asia Studies*, No. 59 (6), September 2007, pp. 913-936.

of Austria is typical of a successful oppression (or indoctrination) by the territorial state, and it may signal that in the long run, the deviant language and ethnicity will disappear through a total assimilation into the dominant majority of the "nation-state"[18].

In the southern part of the river boundary between Finland and Sweden, the two towns of Haparanda and Tornio form one urban agglomeration divided only by the state boundary. However, while Tornio in Finland has an ethnically homogeneous population of Finns, Haparanda is a mixed town, even if the ethnic lines are blurred. Very roughly, one third of the population is ethnically Swedish, one third ethnically Finnish, the rest being the original inhabitants of the Swedish side of the river valley, people whose native language was Finnish for many generations, but who, under the intense pressure of Swedish educational centralism, have more or less gone over to Swedish, especially the youth. Now, with more liberal education, standard Finnish is allowed and even encouraged.

What, then, about identity? In the cases of the Tornio population and the Haparanda Swedes, their loyalty to the respective home state is unquestioned. In fact, even the original Torne Valley population of the Swedish side has been taught to express Swedishness and deny any emotional relation to Finland, even to the degree that some have changed their Finnish family names into names that sound more Swedish. The third group, however, consists of people from Finland who have settled on the Swedish side. With their behaviour, they express an allegiance to Finland in terms of media consumption. There is also a lack of Swedish flags in the areas dominated by Finns. They favour co-operation between the two towns, while the Swedish population, who usually do not speak Finnish, are more negative to contacts with Tornio[19]. Haparanda has by far the lowest voting participation in municipal elections in Sweden, in spite of the fact that even resident citizens of Finland have voting rights – but a very low rate of participation.

In the Russian-Swedish borderland of the 17th century, religion – Lutheran or Orthodox – was usually a stronger marker of "nationality" (Swedish or Russian) than language, but minority groups of Finno-Ugric speech and Orthodox religion were seen as possible cases of "re-conversion" into Lutheranism, while Russian speakers were regarded as

18 T. Lundén, *On the Boundary. About Humans at the End of Territory*, *op. cit.*, pp. 97-101; T. Lundén, "Language, Ethnicity and Boundary Changes in 20th Century Europe", *op. cit.*; T. Lundén, "The Creation of a Dying Language", in *Folia Scandinavica Posnaniensia*, Vol. 12, Poznań, Wydawnictwo Naukowe UAM, 2011, pp. 143-154.

19 T. Lundén, "Border Agglomerations in the Baltic Area: Obstacles and Possibilities for Local Interaction", in *Geographica Helvetica*, Vol. 1, 2007, pp. 22-32.

aliens within the Swedish realm[20]. In the case of Poland, the political subordination of Polish-speakers during the "long 19th century" and during the Communist authoritarianism meant an association of "Polishness" with the Catholic Church[21], in turn resulting in the "ethnic delimitation" towards Protestant, Jewish, Orthodox and Uniate (Greek Catholic) speakers of native Polish or similar dialects[22]. In Poland, ethnicity is strongly tied to religion, as a consequence, native speakers of Polish with evangelical protestant religions were seen and referred to as "Germans"[23]. At the plebiscite of 1920 about the formerly German Ostpreussen, the Protestant Polish-speaking minority almost by 100 per cent voted for their areas to be included into the German exclave of East Prussia and not into the new Polish state[24]. There is thus a divide in religion, not only in denomination, but also in the social and nation-forming role of the church resulting in the choice of group identity[25].

7. Border Areas of Strict Ethnic Division

Many borderlands of Europe show a more distinct separation of identities, because of ethnic cleansing and/or resettlement of population. In the German-Polish borderlands, the German side has a resident population from many generations back, while on the Polish side, the area was resettled, partly with people from the formerly eastern provinces, annexed by the Soviet Union, while the resident Germans were forced to leave[26]. The attitudes and identification of the two groups differ significantly.

---

20 K. Katajala, "Early Modern People(s) in the Borderlands: Linguistic or Religious Definitions of 'Us' and 'the Other'", in M. Hurd (ed.), *Borderland identities. Territory and Belonging in North, Central and East Europe*, Gdansk, Gondolin, 2006, pp. 331-353.

21 N. Davies, *Heart of Europe: The Past in Poland's Present*, (2nd ed.), New York, Oxford University Press, 2001, p. 300.

22 A. Pasieka, *Seven Ways to God: The Dynamics of Religious Pluralism in Rural Southern Poland*, Dissertation at Martin-Luther-Universität Halle-Wittenberg, Halle, 2012, pp. 72-74.

23 A. Kossert, *Preußen, Deutsche oder Polen? Die Masuren im Spannungsfeld des ethnischen Nationalismus 1870-1956*, Wiesbaden, Harrassowitz, 2002; J. von Wedel, "Good Neighbor Inculturation of Religious Communities in West Pomerania", in *Religion in Eastern Europe* XXX, Vol. 3, August 2010, pp. 31-32.

24 A. Kossert, *Ostpreussen. Geschichte und Mythos*, München, Siedler, 2005, pp. 217.

25 M. Marody, and S. Mandes, "National Identity of Poles", in *National Identity Social Sciences Eastern Europe*, GESIS Leibniz Institute for the Social Sciences, June 2009, pp. 10-12.

26 A. Kossert, *Kalte Heimat. Die Geschichte der deutschen Vertriebenen nach 1945*, München, Siedler, 2008; J. Musekamp, *Zwischen Stettin und Szczecin. Metamorphosen einer Stadt zwischen 1945 und 2005*, Frankfurt (Oder), Inaugural dissertation, Europa-Universität Viadrina, 2008.

When in 1945 the German-Polish boundary was decided, the Rivers Oder/Odra and Neisse/Nysa were used as a border marker, with the exception of the Baltic coast area, where the towns of Stettin/Szczecin and Swinemünde/Świnoujście were left to Poland. The outskirts of Świnoujście and Szczecin directly border Germany. In the time of the GDR, the border was mostly closed. Now there are several border crossings, but until recently few of them were open to car traffic. The urban Polish side meets a recreational seaside area of Germany, and further south a rural Germany is a periphery with high unemployment and a high degree of right extremism, partly with strong anti-Polish sentiments. The Polish side however is used by Germans for shopping, but the main urban centres are very peripheral in Poland, Szczecin's nearest large city being Berlin. In this border area, national identities are clear-cut and strong[27]. There is very little bilingualism, few people speak the language of the other side. Polishness is strongly related to Catholic faith, while the local German area is evangelical but very secular. Cross-border cooperation between groups of the same religion is weak because of historical legacies, but recurring[28]. However, with housing shortage in Szczecin and emptied housing areas on the German side, a number of Polish people have settled, usually commuting back to Poland for work. Their children go to school and day-care centres in Germany and probably develop a double identity, but nationalist groups both in Germany and Poland try to prevent any "mixture" of ethnic identity[29].

In and around the Russian exclave of Kaliningrad, national identities are quite clear and linked to each nation-state[30]. In areas of "double" resettlement, e.g. in the former German East Prussia divided between Poland and the USSR/Russia, two populations live side by side but with

---

27 T. Lundén *et al.*, "Szczecin. A Cross-Border Center of Conflict and Cooperation", in J. Janczak (ed.), *Conflict and Cooperation in Divided Towns and Cities*, Berlin, Logos Verlag, 2009, p. 118; P. Balogh, *Perpetual Borders. German-Polish Cross-Border Contacts in the Szczecin Area*. Stockholm, PhD dissertation, Stockholm University, 2014.

28 J. von Wedel, "Good Neighbor Inculturation of Religious Communities in West Pomerania", *op. cit.*, pp. 31-32.

29 F. Barthel, *Auswirkungen der Immigration polnischer Bürger auf den Landkreis Uecker-Randow am Beispiel Löcknitz im Fokus: Wohnungsmarkt, Siedlungsentwicklung sowie Bildungs- und Erziehungsinrichtungen*, Neubrandenburg, Hochschule Brandenburg Schriftenreihe J, 2010; T. Lundén, and P. Balogh, "Trans-Boundary Movements: The Case of the Polish-German Land Border", in *Cultural Diversity as a Source of Integration and Alienation – Nations, Regions, Organizations*, Szczecin, Uniwersytet Szczecinski, 2011, pp. 33-47.

30 T. Lundén, "Contacts and Identity in Border Communities", in T. Lundén, G. Bergström, and L. L. Nilsson (eds.), *Kaliningrad Identity – Crucial to Democracy and Development in the Baltic Sea Region*, Baltic and East European Studies 12, Södertörn University, Centre for Baltic and East European Studies, 2009, pp. 60-65.

little common history. The pre-war population of southern East Prussia was mainly protestant but Polish-speaking Masurians. In the plebiscites of 1920, they chose to be included into German East Prussia by an overwhelming majority[31]. Their identity was rather Protestant than Polish, and after the area was given to Poland after World War II, most of them chose Germany when given a chance to emigrate[32]. Poland and Lithuania have reciprocal ethnic minorities, and there have been certain conflicts regarding their rights to proper official names etc. The Kaliningrad region and its surroundings is thus an area of three nation-states with a history of mixed and changing ethnical identities. People have moved, by force or voluntarily, and territorial identities have had to change[33].

In the Baltic states of the former USSR, nation-building is under way, leaving (mostly Russian) immigrants since 1945 in the choice of identification with the new state of residence or their "new state of origin", i.e. Russia.

8. Conclusions

There is a wide variety of concepts of identity and spatial behaviour in border areas in north-eastern Europe. Spatial identity and allegiance influence informal cross-border contacts, but in the long run, the causal relation could be the opposite.

By financing cross-border activities, the European Union, Norway and Russia have helped to stimulate and create cross-border activities in Northern Europe. But without local incitement, such actions will die once the money is gone. For this reason, informal cooperation across borders will determine if an effective social, cultural and economic use of the borderland is feasible.

There are many hurdles to informal cross-border cooperation. One is a false type of loyalty to the nation-state, the feeling that contacts with "the other" are irrelevant, or even dangerous. And "the other" could be the neighbouring state, the other ethnic group or the other religion. But with safe belonging in a borderland, such divisions could, and should, be erased.

Ethnic identity is a learning process, but so is territorial identity. Well-treated minorities can combine an ethnic identity with a territorial

31 A. Kossert, *Ostpreussen. Geschichte und Mythos*, *op. cit.*

32 A. Kossert, "Religion versus Ethnicity: A Case Study of Nationalism or How Masuria Became a 'Borderland'", in M. Hurd (ed.), *Borderland identities*, Eslöv and Gdansk, Gondolin, 2006, pp. 313-330.

33 T. Lundén, "The Emergence and Decline of Ethnic Minorities in Baltic Borderlands", in *Limes: Borderland Studies*, Vol. 5, No. 12, 2012, p. 137.

allegiance to the state they live in. But we also need cross-over identities, erasing old nationalistic we-and-them attitudes. And such feelings can start to grow if people in border regions learn to know each other while respecting their differences. This could be the foundation for a lasting bottom-up cross-border cooperation process.

A Continuously Ambivalent Border

Building Cooperation Strategies in the Cross-Border Metropolises of Basel and Geneva

François MOULLÉ & Bernard REITEL

Laboratoire DYRT, University Lille Nord de France, Artois

1. Introduction

The city, as a place of centrality and attraction, seems to be opposed to the concept of international boundary[1] (associated to the idea of separation[2] and understood as a limit of sovereignty between two national territories). However, in recent years, scientific interest has been drawn to cross-border urban areas, especially metropolises. These are often considered as the main anchoring points of the globalization of town systems and only a few of them are located on borders. In 1990, L. Herzog described the Greater San Diego-Tijuana (which is on both sides of the border between the United States and Mexico) as a cross-border metropolis. He emphasized its unique character and the symbiotic relationship that exists between the cities across the border[3]. Other researchers have recently emphasized the ambivalent nature of this configuration[4]. According to those researchers, the metropolitan dimension is only located on one side of the border, meanwhile the cross-border urban agglomeration is mainly characterized by the juxtaposition of two urban systems. The purpose of this article is to try to gain a better understanding of this spatial configuration (the cross-border

1 M. Bassand, *Cités, Villes, métropoles. Le changement irréversible de la ville*, Lausanne, Presses Polytechniques et Universitaires Romandes, 2007, p. 234.

2 B. Reitel *et al.* (dir.), *Villes et frontiers*, Paris, Anthropos, 2002, p. 275.

3 L. Herzog, *Where North meets South. Cities, Spaces and Politics on the US-Mexico Border*. Austin, University of Texas Press, 1990, p. 289; C. Hee, and C. Bae, "Tijuana-San Diego: Globalization and the Transborder Metropolis", in *The Annals of Regional Science*, Vol. 37, No. 3, 2003, pp. 463-477.

4 T. Alegría, *Metrópolis transfronteriza*, revisión de la hipótesis y evidencias de Tijuana, México y San Diego, Porrula Miguel Angel, 2009, p. 396.

metropolis) from the examination of two urban areas located in Europe: Basel and Geneva. Several studies have demonstrated the importance of cross-border urban situations in Europe and the existence of cross-border settlements. Among them, some undoubtedly have a metropolitan dimension[5]. We want to show that Basel and Geneva are following these characteristics. In this context, it is necessary to wonder what a cross-border dimension means combined with metropolitan functions. Two hypotheses can be proposed to define a cross-border metropolis:

- A cross-border metropolis is a border or cross-border conurbation with metropolitan functions that enables the organization of a large regional area astride on several national territories;
- On the other hand, a cross-border metropolis is a spatial construction driven by a political project which notably aims to embed the cross-border dimension and to increase the readability of this configuration at different scales.

In short, this chapter aims at questioning the impact of the border on a metropolitan area. Does ambivalence (which is one of the characteristics of the border) work as an asset for agglomerations to join the ranks of world metropolises?

We will first clarify the conceptual and theoretical framework by describing the characteristics of cross-border metropolises. In a second phase, we will seek to compare the two cross-border agglomerations of Geneva and Basel by using an analysis grid. Then, in a third phase, we will describe the proposals of public bodies to elaborate a project of cross-border metropolises.

2. A Cross-Border Metropolis: a Place with Metropolitan Functions that Organizes a Cross-Border Region

In this first part, we will establish the criteria that describe what we call a cross-border metropolis: having the functions of a metropolis and displaying a cross-border dimension[6].

---

5 Espon 1.4.3, *Study on Urban Functions*, Brussels, European Commission, 2007; C. Vandermotten, *Les aires urbaines transfrontalières en Europe*, Mosella 32 (1-4), 2010, p. 37-47; Espon-Metroborder, *Cross-border polycentric metropolitan regions*, Brussels, 2010, European Commission http://www.espon.eu/export/sites/default/Documents/Projects/TargetedAnalyses/METROBORDER/METROBORDER_-_Final_Report_-_29_DEC_2010.

6 C. Sohn, "The Border as a Resource in the Global Urban Space: A Contribution to the Cross-Border Metropolis Hypothesis", in *Journal of Borderlands Studies*, 2013.

In a world organized by flows and networks[7], metropolises are considered as world class centralities[8], transaction centres whose relations with other metropolises are more intense than those held with their neighbouring areas[9]. The metropolis is defined as "an urban system whose configuration makes it possible to access a level of excellence in the world"[10]. Many attempts have been made to define the global cities[11], including the recent work of the GaWC network[12]. In all these analyses, the main question lies on the general criteria that distinguish metropolises from other urban spaces. The size criterion is not sufficient to describe the metropolis, although a minimum threshold is implicitly required to qualify for the term *metropolis*. Other studies have been conducted at the European level. The pioneering work of Roger Brunet[13] within the GIP Reclus did not refer explicitly to metropolises, but it did serve as a basis to get an overall picture of the European system of cities. The analysis provided by R. Brunet helped to determine criteria that establish an urban hierarchy from the comparison of European urban agglomerations. Other studies followed[14]. More recently, several Espon[15] programs[16] on urban functions were initiated. One of them allowed the identification of the Metropolitan European Growth Area (MEGAs) on the basis of five criteria (population, transport, industry, knowledge, decision-making capacity). In this typology, size is an essential criterion, even if combined with other characteristics. We

---

7 M. Castells, *La société en réseaux. L'ère de l'information*, Paris, Fayard, 1996, p. 613.

8 M. Bassand, *Cités, Villes, métropoles. Le changement irréversible de la ville*, *op. cit.*

9 O. Dollfus, *La mondialisation*, Paris, Presses de Sciences Po, 1997, p. 166; L. Halbert, *L'avantage métropolitain*, Paris, PUF, 2010, p. 143.

10 J. Lévy, "Paris métropolitains, réseaux et territoires de l'espace parisien", in J. P. Leresche, D. Joye, and M. Bassand (dir.), *Interdépendances mondiales et implications lémaniques*, Genève, Georg, 1995, p. 58.

11 J. Friedmann, "The World cities hypothesis", in *Development and Change*, Vol. 17, No. 1, 1986, pp. 69-84; S. Sassen, *La ville globale*, New York, London, Tokyo, Paris, Descartes & Cie, 1996, p. 530; P. J. Taylor, *World city network. A global urban analysis*, London, Routledge, 2004, p. 241; G. Antier, *Les stratégies des grandes métropoles. Enjeux, pouvoirs et aménagement*, Paris, Armand Colin, 2005, p. 263.

12 GaWC. *Globalization and World Cities Research Network*. http://www.lboro.ac.uk/gawc/ [accessed on 11th July 2012].

13 R. Brunet, *Les villes européennes*, Montpellier, GIP RECLUS, rapport pour la DATAR, 1989, p. 79.

14 K. Kunzmann, and M. Wegener, *The Pattern of Urbanisation in Western Europe 1960-1990*, Dortmund, report for the European Communities, 1991, p. 351; N. Cattan *et al.*, *Le système des villes européennes*, Paris, Economica, 1994, p. 201; C. Rozenblat, and P. Cicille, *Les villes européennes: analyse comparative*, Paris, La documentation française, 2003, p. 94.

15 European Spatial Planning Observatory Network.

16 Espon 1.1.1, *Potentials for polycentric development in Europe*, Brussels, 2005, p. 348.

have therefore chosen to build upon the work conducted by Blotevogel and Danielzyk[17], who consider that size is not a sufficient criterion to qualify as a metropolis. According to them, four dimensions are necessary in order to consider a city as a metropolis. A metropolis is first a place of innovation and creation; this capacity for innovation is due to the existence of an atmosphere, open to scientific as well as cultural and artistic creativity[18]. Second, the metropolis is a pole with a concentration of economic powers (headquarters of multinational firms). Third, it is an articulation and circulation node (Gateway). Finally, it has a symbolic power, a name and emblems that have a global impact. This last criterion is particularly difficult to identify because it implies more than three others. However, it reflects the idea that a metropolis is also a place that is meaningful in the world-system and shows that the functional criteria are not sufficient to define a metropolis. These four dimensions have the advantage of being sufficiently diverse and at the same time limited enough to enable an effective comparison between Basel and Geneva. The distinction between metropolises and other urban agglomerations is based on the undoubtedly global character of its many functions.

A metropolis is also a political construction[19]. Indeed, many agglomerations have the ambition to project into the world on both the functional and symbolic fields, and take their inspiration from discourses and methods used by public authorities in charge of metropolises considered as absolutely essential. Even if it is often locally inspired, the metropolitan project is also supported by the States in the framework of their territorial development policy. The metropolitan project has three components. It is a social construction, in the sense that it aims at creating a synergy between the key actors around common objectives. It is a spatial construction designed to create consistency in terms of development on what is considered the "metropolitan territory". Finally, it is a political construction that requires the creation of a governance system on a geographical basis[20]. The metropolitan project is often

---

17 H. H. Blotevogel, and R. Danielzyk, "Leistungen und Funktionen von Metropolregione", in J. Knieling, *Metropolregionen. Innovation, Wettbewerb, Handlungsfähigkeit*, Hannover, ARL, 2009, pp. 22-29.

18 R. Florida, *The Rise of the Creative Class... and How It's Transforming Work, Leisure, Community and Everyday Life*, New York, Basic Books, 2002, p. 434.

19 P. Claval, and A. L. Sanguin, *Métropolisation et politique*, Paris, L'Harmattan, 1997, p. 316; N. Brenner, "Metropolitan Institutional Reform and the Rescaling of State Space in Contemporary Western Europe", in *European Urban and Regional Studies*, Vol. 10, No. 4, 2003, p. 297-324; B. Dolez, and D. Paris (dir.), *Métropoles en construction. Territoires, politiques et processus*, Paris, L'Harmattan "logiques politiques", 2004, p. 294.

20 B. Jouve, *La gouvernance urbaine en question*, Paris, Elsevier, 2003, p. 124.

complex as it is conceived both to provide an agglomeration with the necessary infrastructure to generate economic momentum and to increase its visibility at the international level. Due to the high degree of political fragmentation, coordination of actors with different orientations is often an important challenge in metropolitan areas. The analysis of the metropolitan project is not based on the four criteria described above, but on the ambitions, namely the ability, of public bodies to insert the urban area into a network of cities in the world system in order to bring significant profit.

Few authors question the metropolitan space functions. Moreover, the boundary of the agglomeration is not really questioned. Yet, this issue is critical for cross-border agglomerations. Indeed, are the functions describing the metropolis similarly distributed in the various national components of the agglomeration? Or, are they concentrated in a single part? In other words, does the border play a separation role in the distribution of economic functions within the urban area? The cross-border issue is undeniably included in the metropolitan dimension. We will consider the reality of the spatial organization of the cross-border metropolis through different forms of integration[21].

The border is a geopolitical object whose structural effects on the geographic area have been repeatedly mentioned. Historical and cultural heritage deeply marks the imaginary and conditions the relations in border areas. This object is a "territorial construction that puts the distance in the vicinity"[22]. It has properties[23] that affect cross-border dynamics both for the exchanges and the discharges. However, many key actors tap into the gap created by the juxtaposition of distinct legal and institutional realities, including those areas where the border is not appeased. The border becomes a source of profit, while States seek to filter and regulate the flows. Several authors have shown the existence of specific interactions across the border whose intensity reflects cross-border integration[24]. These interactions can lead to the

21 E. Brunet-Jailly, "NAFTA and Cross-Border Relations in Niagara, Detroit and Vancouver", in *Journal of Borderlands Studies*, Vol. 21, No. 2, 2006, pp. 1-19.

22 C. Arbaret-Schultz, "La question du continu et du discontinu à l'épreuve de la dimension technique des sociétés", in F. Alexandre, and A. Génin, *Continu et discontinu dans l'espace géographique*, Tours, Presses Universitaires François Rabelais, 2008, pp. 409-416.

23 C. Raffestin, *Eléments pour une théorie de la frontière*, Diogène, avril-juin 1986, pp. 3-21.

24 R. Ratti, and S. Reichman (eds.), *Theory and practice of transborder cooperation*, Basel, Helbing & Lichtenhahn, 1993, p. 376; O. J. Martinez, *Border People. Life and Society in the US-Mexico Borderlands*, Tucson, The Arizona University Press, 1994, p. 353.

existence of complementary issues that allow us to speak of cross-border regions. But this concept has another dimension: a multidimensional institutional cooperation between the public authorities on both sides of the border, the establishment of joint initiatives and, where appropriate, the establishment of a formal framework for cooperation[25]. As Van Houtum[26] recalls, the cross-border context may be analysed through three approaches. The first approach (the functional one) examines the socio-economic flows and the more or less important role of the barrier effect. The second approach (the institutional one) analyses political and legal frameworks. Finally, a third approach offers a more anthropological dimension by integrating cultural and symbolic dimensions (*people's approach*). We propose to develop an analysis grid of the cross-border dimension from these three approaches by adding the urban morphology element that is appropriate to our two case studies. Continuity of built fabric on both sides of the border implies the existence of perennial links across the border[27].

The integration of territories[28] helps to assess the degree of higher or lower cut-off of the border. Thus, a cross-border metropolis may look like a *built-up area*, even if this feature is not always verified in megapolises. Megapolises look more like *functional spaces*, consisting of one or several morphological agglomerations and places with intense interactions. The political dimension is a component that relies on the degree of integration with, on the one hand, a metropolitan project mobilizing key actors of the cross-border metropolis and with, on the other hand, the existence of a more or less developed institutional framework.

What makes cross-border metropolis different from other metropolises? First, the meeting between the metropolis and the border generates innovation, including new urban forms[29]. While institutional fragmentation

25 B. Wassenberg, *Vers une eurorégion? La coopération transfrontalière franco-germano-suisse dans l'espace du Rhin supérieur de 1975 à 2000*, Bern, Peter Lang, 2007, p. 487.

26 H. van Houtum, "An Overview of European Geographical Research on Borders and Border Regions", in *Journal of Borderlands Studies*, Vol. 15, No. 1, 2000, pp. 57-83.

27 B. Reitel, "L'agglomération transfrontalière: un objet géographique, produit de relations singulières entre continu et discontinu. L'exemple de l'agglomération trinationale de Bâle", in F. Alexandre, and A. Génin, *Continu et discontinu dans l'espace géographique*, Tours, Presses Universitaires François Rabelais, 2008, pp. 311-326.

28 B. Reitel, "Are Cross-Border Urban Spaces European Integration Patterns at Local Level? Analysis of the Management of Urban Areas astride the 'French Border'", in P. Bauer, and M. Darley (eds.), *Borders of the European Union: Strategies of Crossing and Resistance*, Prague, CEFRES, 2007, p. 148-175.

29 F. Moullé, "L'agglomération transfrontalière genevoise: acteurs, stratégies et fonctions internationales", in B. Reitel *et al.*, *Villes et Frontières, Anthropos-Economica*, 2002, pp. 114-123.

characterizes these configurations as in the case of other metropolises, the presence of boundaries introduces a specific context. Indeed, in the framework of cross-border areas, no key public body has long seemed legitimate to think and to bear the destiny of an agglomeration or an urban area as a whole, while in all agglomerations located in a single national country, the State alone assumes that legitimacy. However, not all cross-border metropolises are the same and a comparative approach is therefore needed. One has to combine the four criteria to measure the cross-border area with the four dimensions of the metropolis in order to create an evaluation grid. The border is considered on the one hand as a barrier or an obstacle. On the other hand, we consider the connecting function of a border or, in others words, the existence of a "cross-border reality". The various proposed criteria presented in the form of a grid do not pretend to be exhaustive, but they allow us to establish a comparison between the two scenarios.

3. Basel and Geneva: Two Metropolises and their Cross-Border Areas

The cross-border dimension of agglomerations gives Basel and Geneva an undeniable originality in Europe compared to agglomerations of similar size. A complex and unique urban and political reality is produced.

The metropolitan nature of Basel and Geneva is questioned by the four parameters proposed by Blotevogel and Danielzyk (*ibid.*), i.e.: the *economic pole*, the *communication node*, the *innovation centre* and the *symbolic place*. These parameters must be taken into account along the border or cross-border dimensions of each of them. According to the degree of their cross-border dimension, metropolises can be described and compared.

Basel and Geneva are primarily two cities that are among those with the highest number of inhabitants in Switzerland. They are identified as the major metropolises of the country along with Zurich. Both are at the heart of cross-border agglomerations, urban areas characterized by a morphological continuity of the built fabric astride one or several international borders. Both agglomerations have a strong institutional fragmentation, despite two very different locations: the city of Basel borders the French and German boundary, while Geneva is a few kilometres away from the French border. Basel is a tri-national agglomeration; Geneva is bi-national. In addition, another feature of these communities lies in their Trans-cantonal character: the suburbs of Geneva also include the Canton of Vaud; Basel covers the cantons of Basel-campaign, Aargau and Solothurn. The Cantons of Geneva and Basel-City, despite their small size, are similar to city-states that have a strong political autonomy[30]. This configuration of isolation has made them soon aware of their fragility in an environment that could sometimes appear hostile, as some elements show. The presence of the Kingdom of France on the outskirts of Basel has resulted in the construction of the citadel of Huningue. Carouge, which is currently a municipality in the suburbs of Geneva, was a possession of the Kingdom of Piedmont-Sardinia. It became a royal town in 1786. Basel was divided into two half-cantons in 1833 after a revolt of rural communities against the powers of the Basel middle class. Geneva is still considered today as a city-republic.

Table 1. Criteria Proposal for a Cross-border Metropolis. Test for Basel and Geneva

	The Basel Agglomeration		**The Geneva Agglomeration**	
	Border dimension	**Cross-border dimension**	**Border dimension**	**Cross-border dimension**
Metropolitan properties				
Economic pole	Pharmaceutical companies, exhibitions (watchmaking, art), international organizations (Bank for International Settlements) international transport companies,		Banks, insurances, headquarters for companies and NGOs	Precision industries, pharmacy, electronics, logistics
Communication node	Transport intermodal pole (river, road, rail), railway and motorway node, Swiss entry	Airport, cooperation between the river ports.	Railway to be fostered	aerial, road, numerical

30 B. Reitel, "Are Cross-Border Urban Spaces European Integration Patterns at Local Level? Analysis of the Management of Urban Areas astride the 'French Border'", *op. cit.*

	The Basel Agglomeration		The Geneva Agglomeration	
	Border dimension	**Cross-border dimension**	**Border dimension**	**Cross-border dimension**
Innovation centre	Life sciences and pharmaceutical products		Higher tertiary	Industrial, CERN
Symbolic place	Humanism, contemporary architecture and art	Contemporary architecture and art	Reform, humanism, internationalization, art	
Cross-border integration level				
Morphological agglomeration (a)		Type 2		Type 2
Functional space	Life spaces	Cross-border employment area		Cross-border employment area, life spaces in cross-border areas
Metropolitan project	Metropolitan region, territorial marketing and economic promotion	2 different projects: one initiated by the Basel Canton-city (urban planning and structuring of the agglomeration), another by the economic world (think tank and lobbying), one event project: architecture and town planning international exhibition (IBA 2020)	Worldwide economic promotion	1 agglomeration project began by the Canton of Geneva and the Regional Committee French-Geneva (CRFG): urban planning and structuring the agglomeration. Economic promotion at the regional level

	The Basel Agglomeration		**The Geneva Agglomeration**	
	Border dimension	**Cross-border dimension**	**Border dimension**	**Cross-border dimension**
Institutional framework		Basel Tri-national Eurodistrict (ETB): Association with political structure, metroBasel, exchanges platform and think tank.		Regional Committee French-Geneva (CRFG)

(a) Metropolis constituted by a morphological cross-border agglomeration that includes a secondary pole from the other side of the border and a cross-border functional surface. Espon 1.4.3, *Study on Urban Functions*, 2007.

Both agglomerations have a comparable size, but their demographic dynamics undoubtedly distinguish them. The agglomeration of Geneva has had, for several decades, a high population growth, while the Basel agglomeration has recorded a relative stagnation since the 1970s. Geneva has supplanted Basel both as a city and as an agglomeration with the highest number of inhabitants within the Confederation after Zurich. However, the analysis of the indicators shows that the two agglomerations fulfil all the criteria in order to be classified as metropolises (see Table 1).

Geneva and Basel can be called metropolises in the sense that the two places really weight on world scale. Thus, the two metropolises are both real economic poles at the regional level, communication nodes at all levels, innovation centres in the industry and in the services activities, and they also perform symbolic functions[31].

First, their economic centrality makes no doubt at a global level, even if each city has its own characteristics. Basel is the headquarter of two major global pharmaceutical companies. Geneva is the headquarter of companies with more modest dimensions (precision industries) and is especially a world-class financial centre. It also hosts many international organizations. In this field, Basel pales compared to Geneva, despite the presence of the Bank for International Settlements (BIS).

[31] H. H. Blotevogel, and R. Danielzyk, "Leistungen und Funktionen von Metropolregione", *op. cit.*

This concentration of economic power and international organizations is remarkable for two agglomerations of modest size[32].

Basel and Geneva are also two main entry gates to Switzerland. Geneva airport allows the compensation for its relative isolation on the rail and road map, even if it is small compared with others worldwide. Basel enjoys a convenient location on the Rhine corridor and forms one of its main nodes. The city is an unavoidable point of passage between Northern Europe and Southern Europe. The agglomeration appears as a node that connects several modes of transport, inland waterways, railways and road: Basel is a multi-modal and intermodal platform on the Rhine route[33]. This communication function cannot be separated from their economic dynamism.

The two agglomerations are also scientific research poles. Basel is a major centre of innovation in life sciences. This specialization is mainly focused in the pharmaceutical companies. Geneva hosts the CERN that is a public research centre specialized in nuclear physics research. In both cases, the cities are also important centres of artistic creation and places with exhibition venues.

Finally, the two cities have a real symbolic dimension. The cultural and humanistic influence of Geneva is symbolized by Calvinism, the presence of the Red Cross, the Office of the United Nations and by many international NGOs activities. The humanist dimension is also present in Basel, while it is widely associated to Contemporary Art and to architectural and urban innovation. In both cases, these cities are involved in organizing major international events: art fair, watches fair in Basel, motor show in Geneva.

The metropolitan attributes of these two poles of modest dimension are, in addition, relatively concentrated in the urban area. In this context, one can ask if the four metropolitan properties have a cross-border dimension, i.e. if these functions are located throughout the cross-border area or whether on the contrary they are concentrated in one national part of the agglomeration. An examination of their degree of cross-border integration will complete the analysis.

In general, metropolitan functions only exceed the border in exceptional circumstances. However, cross-border integration is real

---

32 C. Rozenblat, and P. Cicille, *Les villes européennes: analyse comparative*, *op. cit.*; BBSR. *Metropolräume in Europa*, 2011, http://www.bbsr.bund.de/nn_21272/BBSR/DE/Veroeffentlichungen/BerichteKompakt/2010/DL__4__2010,templateId=raw,property=publicationFile.pdf/DL_4_2010.pdf [accessed on 14th June 2011].

33 A. Beyer, "Nœuds de transport et frontières. L'invention de la métropole bâloise", in *Annales de géographie*, Vol. 657, No. 5, 2007, pp. 451-469.

because Basel and Geneva organize around them genuine urban regions[34], which is reminiscent of the concept of growth pole[35].

Both in Basel and Geneva, economic power is relatively concentrated: the headquarters of companies are located on Swiss territory. A location of the companies on both sides of the border reflects a specialization: decision and research are located in Switzerland, production and logistics in France or in Germany. Other metropolitan functions show a more symmetrical distribution. In some cases, the division of functions is the subject of agreements between the neighboured countries. The bi-national status of EuroAirport (located on the French territory due to lack of land on Swiss territory, but connected by a way without customs to the Swiss territory) has been the subject of an international convention in 1949. This airport is a symbol of cross-border cooperation and its attraction area spreads over three national territories. But it is also synonymous of conflict. The French consider that the nuisance mainly affects their territory[36]. Transport companies have preferred to settle in the "Swiss area" of the airport to take advantage of more favourable conditions for tax purposes and in terms of labour law[37]. This situation, which was called into question by a decision of the Supreme Court stating that the institutions are on French soil and therefore must apply French law, has created great uncertainty for several months. Faced with the threat of several companies to leave the place, the French and the Swiss authorities signed an agreement recognizing the presence of the companies on "Swiss soil". The CERN, attracting research from the whole world, also has a cross-border location.

Another form of integration is possible: the cross-border peripheral public bodies can exploit the image of the main pole to develop their urban projects. The architecture and art associated with the image of the city of Basel have become elements that are shared in the whole cross-border urban area. This results from the willingness of actors of the peripheral cross-border areas to develop positioning strategies that target the metropolitan pole[38]. Basel is home of many museums including several specialized

---

34 C. Sohn, B. Reitel, and O. Walther, "Metropolitan Cross-Border Integration in Europe: Evidence from Luxembourg, Basel and Geneva", in *Environment and Planning C – Government and Policy*, Vol. 27, No. 5, 2009, pp. 922-939.

35 F. Perroux, "Note sur la ville considérée comme pôle de développement et comme foyer de progrès", in *Tiers-Monde*, Vol. 32, No. 8, 1967, p. 1147-1158.

36 M. Flitner, *Lärm an der Grenze. Fluglärm und Umweltgerechtigkeit am Beispiel des binationalen Flughafens Basel-Mulhouse*, Stuttgart, Franz Steiner, 2007, p. 283.

37 O. Walther, and B. Reitel, "Cross-Border Policy Networks in the Basel Region: the Effect of National Borders and Brokerage Roles", in *Space & Polity*, Vol. 17, No. 2, 2013, pp. 217-236.

38 B. Reitel, "Are Cross-Border Urban Spaces European Integration Patterns at Local Level? Analysis of the Management of Urban Areas astride the 'French Border'", *op. cit.*

establishments in contemporary art (Art Basel). Basel organizes an annual contemporary art fair whose reputation is global. Moreover, after several decades, construction projects have been entrusted to renowned architects who build emblematic buildings. This trend increased in the 1990s and led to the emergence of Herzog & De Meuron, one of the most famous international architecture firms in the world. In Weil am Rhein (a German city close to Basel), the Vitra company specialized in the manufacturing of furniture design has created a museum, which was built by Frank Gehry in 1989. It has also called on other several famous architects, like the Londoner Zaha Hadid, to erect some buildings for the company. A similar approach was initiated in the French city of Saint-Louis by the municipality that has transformed a neglected industrial building into an exhibition space for contemporary art. This project was entrusted to the Parisian architect Jean-Michel Vilmotte and intended to be part of a cross-border urban landscape marked by both contemporary art and innovative architecture dimensions (http://www.museefernetbranca.org/page00010033.html). Compared to Basel institutions, the museum Fernet-Branca of Saint-Louis remains modest in size. The cross-border dimension is not accidental: it comes from a transposition plan of the image representing the central pole towards the periphery. A cross-border urban identity seems to be gradually emerging.

Geneva and Basel are each the core of functional areas straddling several national territories thanks to the attractiveness of the employment pole of the Swiss part. The urban sprawl and the concentration of jobs logically cause a high mobility: the phenomenon is common to all metropolises and at the same time an original feature in a cross-border context where the planning of transport networks is a recent and unfinished reflection[39]. In both cases, the intensity of institutional cross-border cooperation and a continuous dialogue between public bodies across the borders facilitate the emergence of a metropolitan project. The situation seems relatively clear in Geneva where only one project was designed. On the contrary, two projects have been proposed in Basel: one by public bodies and the other one by private actors, which lead to a feeling of competition.

Basel and Geneva fulfil the criteria to qualify as metropolises, but the functions that characterize them are mainly located on the Swiss part of the agglomerations. However, the cross-border dimension lays on the high rate of commuters and other flows. The presence of an international border between two States gives the main pole of the metropolis the possibility to

---

[39] C. Jemelin, and V. Kaufmann, *Le maintien du tramway envers et contre tout: politique des transports et démocratie directe à Bâle*, Flux, 2008, pp. 72-73, pp. 92-107; A. Beyer, and B. Reitel, "Les systèmes de transports publics comme mesure de l'intégration transfrontalière", in *Géo-Regards*, Vol. 4, 2011. p. 77-94.

access easily to an international level in a context of proximity. Compared to agglomerations of similar size, but spread over a single national country, these features introduce an additional degree of complexity. In other words, the cross-border metropolis has similar characteristics in terms of governance to those of very large cities. One can in this case talk about cross-border metropolitan regions characterized by a complexity that is similar to that described for megacities: the metropolitan centre primarily situated on one side of the border organizes and influences neighbouring border areas. However, the metropolitan functions are not shared across the cross-border agglomeration, with a few exceptions. When this is the case, these functions are related to bilateral legal agreements between States and concerns specific equipment or infrastructure essential to the operation and influence of the metropolis. Another possibility also exists: an actor in the cross-border periphery may develop a project in relation to the speciality of the metropolitan centre. This approach, similar to an integration strategy, cannot be easily reproduced because of the limited resources available to these urban and cross-border peripheries. The two metropolises are considered as spatial patterns as well as political constructions.

4. The Cross-Border Metropolis: a Political Construction

In the assessment of the cross-border dimension of a metropolis, two criteria are particularly important for the awareness of current issues and the ability to imagine the city in the future. There is, on the one hand, the existence of political bodies of cross-border cooperation with potential performance based largely on the quality of relations. This cooperation can take many forms: it can be informal or otherwise be institutionalized with the existence of formal committees composed of respected and recognized personalities by the acting parties. On the other hand, there is the ability within the cross-border authorities to think of an urban project and to put it into place, which implies the existence of a common and shared vision to provide answers for the future of the agglomeration.

The cross-border dimension undoubtedly disrupts the organization of the urban area and its management[40]. The cross-border urban project has several objectives. The first one is to imagine the whole urban area in the future, in ten or twenty years: spatial balance and fluidity of mobility are central to the project. The second objective is to reduce the barrier effects of the border in order to take full advantage of the resources provided by the existence and identification of the differences.

[40] B. Jouve, *Urbanisme et frontières. Le cas franco-genevois*, Paris, L'Harmattan, 1994, p. 269; B. Reitel *et al.* (dir.), *Villes et frontiers*, *op. cit.*

These metropolitan projects are both familiar and unique. What makes them familiar lies in the vocabulary used and the issues presented. The general pattern of an urban project gives at first sight the feeling to be easily applied from a metropolis to another. Sustainable developments, urban sprawl, economic dynamism and the aging of the population appear as key issues. However, and this is especially true for Geneva and Basel, singularities exist: they are a result mainly of the cross-border dimension that demands specific guidance.

The first issue is to make territories, considered as peripheries by the central State and therefore supposedly ignored at the institutional level, live together. In order to imagine a balance between a national feeling belonging and a lived cross-border area one requires a temporal investment of a collective imagination. A balance must be found taking into account the existence of strong disparities, wages, cost of land, without jeopardizing the border activities and cross-border interactions. *Living together* is also a mutual understanding to jointly find solutions to problems. The Regional French Geneva Committee is particularly exemplary in this regard after now four decades expertise including many fiscal arrangements and access to care. In Basel, the question of living together has been asked since the 1960s thanks to the Regio Basiliensis founded in 1963. This association emanating from civil society aims to bring out a "territory" without borders. Many studies and demonstrations have been initiated to consider cross-border arrangements. The process of Tri-national Agglomeration of Basel[41] initiated in 1995 has facilitated discussions between cross-border territorial communities, which allowed the emergence of a shared vision across the agglomeration. The main aim was to coordinate the territorial planning. The harmonizing action has allowed them to think in the common future of the agglomeration, in terms of transport networks and landscape organization.

In both cases, projects need time to be materialized. Decisions are in principle taken by the public authorities of each country. This fact puts into question the legitimacy of cross-border cooperation frameworks.

This is the second issue regarding these territories. On both sides of the border, citizens expect to have information from democratic structures. The principle of election creates a link between the individual and the representatives of public structures. Communication tools such as newspapers allow individuals to monitor the evolution of projects. However, there is no formal cross-border democratic structure in either Basel or Geneva. Cross-border bodies are often made of a political assembly that includes elected members appointed by the authorities

[41] ATB, *Concept de développement et projets-clés pour l'Agglomération Trinationale de Bâle*, Bâle, rapport final, 2002, p. 177.

of each territory and a technical committee composed of specialized administrative staff. There is no direct link between citizens and these institutions, which is also often the case for inter-communal structures in France. A third issue hence arises: visibility.

Even in Western Europe, cross-border cooperation is not a priority in government actions and this can be even more the case when power is far from the border. For regional or national powers, cross-border issues concerning outlying areas appear to be a secondary concern. Communication through territorial marketing has emerged as a central theme because the cross-border metropolitan project is not "expected" by citizens and influential public actors are far away. Discourses are often based on the idea of designing concrete projects that can be legible to citizens or the idea of solving problems related to the presence of the border. In both cases, these arguments can be heard by the State and by other public bodies.

A fourth issue concerns the organisational logic of the cross-border area covered by the metropolitan project. The most relevant example for understanding its dynamics is the organization of cross-border public transport. As elsewhere, mobility is considered essential, but the presence of the boundary introduces legal, political and technical differences. The construction of a public transport network is considered as a concrete act of cross-border policy. In addition, the Swiss Confederation implemented in 1999 an agglomeration policy whose aim is to strengthen partnerships between communities in charge of urban areas, and especially to formulate extension policies of transport infrastructure with urban development. In Basel and Geneva, transportation projects are designed to ensure a better integration of border areas at two levels: the built-up fabric and the whole urban area[42]. The introduction of faster and more efficient modes of transport is seen as an answer to the difficulties of mobility across borders. Policies also aim to enhance the use of public transport over individual cars. In fact, the border often reveals a gap in the modal split of transport. Extending links in the cross-border area is a way to promote convergence in this distribution to limit the effects of congestion on the border and to avoid exerting heavy pollution on the Swiss part of the agglomerations, where public authorities pay greater attention to quality of life[43]. On both sides of the border, however, the issues are not considered in the same way. On the Swiss side, network

---

[42] Eurodictrict Trinational de Bâle. *Un avenir à trois: stratégie de développement 2006-2020*, Tome 2: La stratégie de développement, Bâle, 2006, p. 90; Kanton Basel-Stadt. *Gesamtverkehrsmodell der Region Basel*, Basel, 2007, p. 34.

[43] C. Jemelin, and V. Kaufmann, *Le maintien du tramway envers et contre tout: politique des transports et démocratie directe à Bâle*, *op. cit.*; A. Beyer, and B. Reitel, "Les systèmes de transports publics comme mesure de l'intégration transfrontalière", *op. cit.*

expansion consolidates access to the heart of the agglomeration and its centrality. In addition, the public transport network is seen as a tool that promotes densification, which brings value to the metropolitan dimension. In Geneva, until recently, 75% of land reserves of building zones were poorly or not served by public transport. Creating common rapid axes of public transport is a way to increase densification in the first suburb belt. In cross-border peripheries, extensions are seen as ways to reinforce centralities: the objective is to benefit from dynamics by developing links with the metropolitan core. Projects in progress are subject to reluctance or resistance. However, spatial integration across networks is often seen as a threat: that of being frequently confronted with otherness and difference without having desired it. Temporalities are not the same in both agglomerations. In Basel, a collective transport system at regional level, the S-Bahn type was designed in the 1980s. Thanks to the Regio Basiliensis the cross-border dimension was integrated upstream in the reflection. The first transport line was inaugurated in 1997 with Mulhouse in France. The network created includes a branch towards France and three towards Germany. Despite some tensions with France, Basel has a cross-border regional public transport network that authorities are trying to consolidate. Now, current projects focus mainly on urban public transport across the agglomeration. A tram line is being extended to Weil am Rhein in Germany while another line is planned to Saint-Louis in 2016. These projects are part of the development strategy of the agglomeration over a 2020 horizon. The goal is to make the city centre accessible in fifteen minutes from a large part of the cross-border agglomeration in S-Bahn or in urban public transport (tram, bus). Thanks to the extension of these networks, the integration of the German side to the Swiss part of the agglomeration seems nowadays better than that of the French side. In Geneva, the completion of the rail link CEVA (Cornavin – Eaux-Vives – Annemasse) is the flagship project of the agglomeration program. This latter is a missing link to connect different suburbs together through the city centre. Started in December 2011, this construction work is linked to the first release of financial funds from the Swiss Federation, entitled to give assistance to an agglomeration project. This massive construction will impact both sides of the border with the emergence of a real cross-border public transport network on the urban area scale. Each regional train station is designed as an urban node to densify. For the French public bodies, this project is a possibility to have a fast link between Annemasse, the largest French satellite of Geneva, and the centre of Geneva, and to design the Annemasse station as a multimodal node for both the French suburbs and the periurban space. All projects are strengthening the unification of the network even if the management will remain separated.

This project reveals both the importance of the current saturation of individual mobility in the cross-border agglomeration and the challenges of densification both in Geneva and in Annemasse, In addition, stakeholders' analysis is used to highlight the ability of operators to create effective responses such as the price integration. It is interesting to note that the Swiss operator (TPG, Geneva Public Transports) is shareholder of TAC (Annemasse Agglomeration Transport) to 49%. The remaining 51% shares are held by the RATP, which has experience in complex networks in very large agglomerations.

In Basel and Geneva, metropolitan projects have gone beyond a design stage to enter an operational phase. Swiss communities benefit primarily from the federal program to structure Swiss agglomerations. This program allows having significant financial aid that facilitates operational decisions. Confederation contributes in this way to the realization of transport infrastructure on foreign territory within the perimeter border if Swiss municipalities and Cantons can prove a beneficial effect on Swiss territory. In the Basel agglomeration, the extension of the tramway to Weil am Rhein is funded by more than 40% of Swiss funds. The Confederation has recognized Basel and Geneva as two cross-border agglomerations, as a concrete reality based on a perimeter identified by local public bodies. Canton appears as a key entity that plays a role of articulation among the peripheries and that initiates cross-border governance. The cantonal governments have significant executive and legislative powers that French regions do not have. Both Cantons, as communities that exercise leadership over the cross-border region impel, direct, put in relationship and create synergies. Their peripheral location and the character of "city-state" of the two Cantons of Basel-city and Geneva give them a unique status in comparison with the rest of Switzerland. Both Cantons are aware that part of their urban development is and will be with neighbouring foreign territories. The urban area of Basel covers Germany, France and neighbouring cantons including Basel-Country. The area of urban Geneva includes two French departments and the Canton of Vaud. Urban growth leads to pressure in terms of residence, work and cross-border mobility. Mobility and the ability to organize individual and collective transport are considered major issues for cross-border metropolitan projects.

The combination of cross-border dynamics has contradictory effects the metropolitan project is trying to reduce. On the one hand, the extension of cross-border networks supports spatial integration, which is also the result of the existence of joint projects and cross-border governance in which urban Cantons (Basel-city and Geneva) exercise a leadership role. However, the border is and will remain a territorial discontinuity on the legal, fiscal, economic and symbolic dimension. On the other hand, the border exacerbates the contrasts within the peripheral areas. The lack of a real integrated management of the territories necessarily causes differences in the treatment of the cross-border urban spaces. Centres concentrate the wealth while cross-border peripheries, specialised in residence functions, do not have the same financial capacity even if fiscal equalisation systems exist like in Geneva. Moreover, wage differences on both sides of the border introduce a socio-spatial discontinuity between those who participate in cross-border dynamics and those who are from the other side. A cross-border metropolitan project cannot erase all the ambivalences of the border, on the contrary. The risk of a weak political integration remains real, despite the will of coherence initiated by the government at national, regional and local level.

5. Conclusion

The objective of this chapter was to describe some features of what can be considered as cross-border metropolitan regions in Europe based on the two cases of Basel and Geneva. The aim was not to understand the processes that led to the emergence of these entities. A cross-border metropolis is an urban space that concentrates metropolitan functions, where a national component gathers the majority of dimensions; meanwhile it organizes a broad space beyond national borders. In addition, it has a higher degree of complexity than its size would suggest, due to the presence of boundaries. It has thus certain similarities with very large urban areas.

The cross-border metropolis is both a political and a spatial construction held by a system of governance in which a public body provides leadership, but where all the various public authorities are involved in the development of a joint project. This project of cross-border metropolis is both a discourse and a plan whose aim is to make the urban area visible, especially among national governments and European institutions. The effectiveness of the project questions the fact of territorial complexity: the border continues to differentiate the legal frameworks and ways of thoughts.

Basel and Geneva have features that are not based on their cross-border dimension. Public authorities, primarily Cantons, and this is specific of the Swiss Federation, play a fundamental role: they join resources and means to enable them to exercise a leadership at the level of a cross-border urban region. The specific features of national territories have effects on the future shape of the cross-border area.

The cross-border region is characterized by a metropolitan asymmetry. This fact highlights the sustainability of the ambivalent nature of the border; it opens up opportunities and allows access to differences within proximity, while it reveals socio-spatial inequalities that are inherent in metropolitan areas.

Endeavours to Establish EGTCs on Hungary's External Borders

Edit Soós

University of Szeged

1. Introduction

Hungary is situated in the centre of the Carpathian basin and borders with seven countries (Austria, Slovenia, Croatia, Serbia, Romania, Ukraine and Slovakia). It has a relatively large number of border regions in spite of the small size of the country, and the border length is nearly 2,246 km[1]. The north-eastern and southern border areas of Hungary presently comprise the external borders of the EU. Those that are regarded as external borders are the Hungarian-Ukrainian and Hungarian-Serbian border lines, featured by relative economic backwardness and past historical events.

In 2006 the European Union created an opportunity for members to establish EGTCs, bodies with legal personality to facilitate the cooperation of national, local and regional players and the efficient use of Union's resources. Hungary is among the leading countries when it comes to EGTC in the internal border area. Twelve of them are seated in Hungary and four have Hungarian partners. The study analyses the reasons, why the EGTC setups with participation of third countries' local governments on the Union's external borders cannot achieve the approval by national authorities, therefore the registration in Hungary[2].

---

1 J. Rechnitzer, "The Dividing Line: Border", in *Minorities in a Seamless Europe. The Role of Transfrontier Cooperation in Maintaining Ethno-cultural Diversity*, International Conference, Budapest, 15th September 2009, Budapest, Prime Minister's Office, 2010, p. 21.

2 16 EGTCs have at least one Hungarian partner (European Common Future Building EGTC (2012), Gate to Europe EGTC (2012), Bodrogközi EGTC (2012), Pannon EGTC (2012), Novohrad-Nógrád EGTC (2011), Rába-Duna-Vág EGTC (2011), Arrabona EGTC (2011), Abaúj-Abaújban EGTC (2010), Banat Triplex Confinium EGTC (2011), Ister-Granum EGTC (2008), Ung-Tisza-Túr-Sajó EGTC (2009),

2. Development of Organic Regions on the External Borders of Hungary

Borders are the 'scars of History'[3]. In the Carpathian basin, the borders of the states are the aftermath of the peace system put into shape after the Second World War. Borders were rather the symbol of the sovereignty of the states during the last centuries and their dividing role prevailed over the connective one.

Borders cut the historically, geographically, socially, politically homogenous organic space, preventing interactions between the two sides of the border, thus presenting problems for national minorities and for their families living on either side of the border. Concerning the number of ethnic Hungarian minority it is 253,899 (2011) in Serbia (Vojvodina), it is 151,533 (2011) in Ukraine (Transcarpathia), while in Romania (Transylvania) it is 1,238,000 (2012) and in Slovakia 458,467 (2011).

However, the geographical environments, the common history, and the survival of the regional consciousness of the people have an impact on the development of the border region.

The significance of borders was transformed, and the role of cross-border regions changed during the last two decades. Cross-border networks in Hungary's north-eastern and southern border area draw up lines of the historical organic region. The cohesion of the border region is steadily integrated by regional consciousness and the inter-institutional system. While the Hungarian national minority belongs to their nation state spontaneously as citizens, they belong to their kin-state in another way, through their secondary identity. The nation states often see cross-border cooperation as a potential threat to the territorial integrity of a state[4] when it deals with ethnic homogeneity rather than with diversity.

Because of the differing structure of national identities the cultural borders of nations do not overlap the political borders of a state[5]. But the EU makes a new conception of the nation state in every state. In the focus

---

Sajó-Rima EGTC (2013), (Karst-Bodva EGTC (2009), Pons Danubii EGTC (2010), European Urban Knowledge Network (2012), Via Carpatia (2013).

3 Association of European Border Regions, *European Charter for Border and Cross-border Regions. Draft New version*, Gronau, 15th September 2011, p. 4. http://www.aebr.eu/files/publications/110915_Charta_EN_clean.pdf [accessed on 21st August 2013].

4 F. Palermo, "Senior Legal Adviser to the OSCE. High Commissioner on National Minorities", in *Minorities in a Seamless Europe. The role of transfrontier cooperation in Maintaining Ethno-cultural Diversity*, *op. cit.*, p. 12.

5 M. Bakk, *Politikai közösség és identitás [Political Community and Identity]*, Cluj, Korunk Komp-Press Kiadó, 2008, p. 77.

of the new conception is the European citizenship which retains the legal and political framework but draws attention to the sources of the cultural and psychological considerations of national consciousness, dissolving the difference between the two paradigms of political nation and cultural nation[6].

In Central and Eastern Europe the countries had exceptionally closed borders with the West. With the opening of the borders in the 1990s the chance for cooperation became available. During the transformation process, the notion of political nation changed. The borders are not only the symbols of nation-state sovereignty but they also have an impact on the relations between subnational levels. Regarding their functions, the borders do not only separate state and territorial structures, but in certain cases they make connections between them.

The removal of borderlines and the reinforcement of their bridge role can induce considerable development in the border area. Attention is currently centred on cross-border cooperation which can allow regions earlier belonging together to reunite, to create areas of natural attraction, to integrate border regions in a peripheral situation, and to strengthen development and cooperation relations in areas inhabited by national minority (Hungarian) populations in varying ways and to varying extents. For this reason, in Hungary, border cooperation appears as a necessary policy in order to most efficiently exploit the wide spectrum of development opportunities on both sides of the border. The external borders, due to the Schengen acquis, are evidently in a disadvantageous situation. The strategy towards Hungarian population can play a key role in the elaboration of the strategic purposes of cooperation across external borders.

The first steps towards organised cooperation started on Hungary's external borders as early as in the 1990s. Initially, it was done spontaneously, with bottom-up cooperation systems evolving without any formality. Later, principles and legal regulations influenced the already functioning practice. Attention is currently centred on European Grouping of Territorial Cooperation (EGTC). The new tool for cooperation across the borders was approved by the EU in 2006[7].

The EGTC instrument was proposed to overcome the obstacles hindering territorial cooperation in the EU and beyond the EU member states' borders. The EGTCs make it possible to reunite regions earlier belonging together, create areas of natural attraction, integrate border

---

6 G. Csepeli, *et al.* (eds.), *Nemzeti érzés és európai identitás [National Feeling and European Identity]*, Budapest, Arkisz Kiadó, Balassi Kiadó, 2007, p. 82.

7 European Parliament and Council of the European Union, *Regulation (EC) No. 1082/2006 on the European Grouping of Territorial Cooperation (EGTC)*, OJ L 2010 31/7/2006, pp. 19-24.

regions in a peripheral situation, and strengthen development and cooperation relations in areas inhabited by Hungarian populations in varying ways and to varying extents.

Hungary was the first to pass its national act[8], necessary for the establishment of EGTCs. In accordance with this act, an EGTC seated in Hungary as a legal entity has limited responsibility. A potential Hungarian EGTC member can only cooperate with a member that operates on a non-profit basis. At the same time, the establishment of an EGTC requires institutions from at least two EU Member States that are members of the cooperation system.

The character of cross-border cooperation differs depending on whether the cooperation concerns areas along internal or external borders. In particular, practical difficulties arise in the application of the EGTCs in the external border areas, where the cross-border cooperation strategy is very different from that in the internal ones. A decisive role has been given to the European integration in the border area. Cooperation makes an indirect contribution to the mediation of European values (rule of law, market economy and respect of human and minority rights).

The EU-supported operational programmes in the regions along the external borders contribute to the cross-border parts of IPA and ENPI (see Table 1). Their activities in many cases go further beyond the implementation of cooperation programmes and projects in the framework of Hungary's relations to Hungarian ethnic minorities living over the border.

But cooperation faces many barriers, including practical and political obstacles. The increasing or decreasing significance of the dividing border line of a state in many cases depends on the intentions of state authorities. Cross-border relations were strongly influenced by the relationship between two neighbouring countries. Border region activities were understood as the relationship between one state's internal political and economic disposition to that of the other[9]. In the cross-border area the local and regional actors' fundraising ability is restrained without support of the central authorities. (Although there are some exceptions, since local governmental and economic actors (enterprises) can have their own resources.)

---

[8] Hungarian Parliament, *Act XCIX. of 2007 on the European Grouping of Territorial Cooperation.* http://www.complex.hu/kzldat/t0700099.htm/t0700099.htm [accessed on 21st August 2013].

[9] W. Aschauer, "Határokon túlnyúló magatartás és a határhelyzet tudatossága az osztrák-magyar határ térségében" [Cross-Border Behavior and the Awareness of the Frontier Situation in the Austro-Hungarian Border Region], in *Tér és Társadalom (Space & Society)*, Vol. 9, No. 3-4, 1995, pp. 157-178.

Table 1. Cross-Border Financial Assistance, 2007-2013

Programme	**Source (€)**
Hungary-Serbia	50,111,383
Hungary-Ukraine-Slovakia-Romania	68,640,000
Altogether	**118,751,383**

Source: http://www.nemzetpolitika.gov.hu/index.php?main_category=2&action=view_item&item=396.

3. International Regulations of Cross-Border Cooperations

There are different challenges at European (transnational), national and subnational levels to cross-border cooperation. At European level there is only one document that makes an attempt to create a comprehensive regulation on cross-border cooperation systems, namely the European Outline Convention on Transfrontier Cooperation[10], passed by the Council of Europe in 1980. The Council of Europe has played a significant part in dismantling barriers to local and international cooperation as well as in strengthening cooperation across borders.

The Convention must also create frameworks of bilateral and multilateral agreements; it can be applied to the local and territorial relations of the ratifying member states.

The Outline Convention increases the role of local self-governments and territorial authorities in creating relations across borders. Therefore, in the member states of the Convention, it is necessary to draft national regulations compatible to the Outline Convention that guarantee political power to communities and self-governments and provide suitable financial resources to create cross-border cooperation systems. For this reason, the central governments of states have to transfer the necessary public legal power not only to local, but also to territorial units so that they can actively contribute to the creation of EGTCs[11].

Hungary signed the Convention on 6th April 1992, and it was announced together with Act XXV of 1997. Nevertheless, the actual creation and operation of cooperation systems largely depends on the political attitude and legal system of the neighbouring countries (see Table 2).

---

10 Council of Europe, *European Outline Convention on Transfrontier Co-operation between Territorial Communities or Authorities*, CETS No. 106, http://conventions.coe.int/Treaty/Commun/ChercheSig.asp?NT=106&CM=&DF=&CL=ENG [accessed on 2nd August 2013].

11 V. Forgácsné Orosz, "*A Madridi Egyezmény végrehajtása Magyarországon*" [The Implementation of the European Outline Convention in Hungary], in *Magyar Közigazgatás (Hungarian Public Administration)*, Vol. 52, No. 2, 2002, pp. 96-98.

Table 2. The Legal Framework of EGTC Setups

Community Law	EC regulations (with supplement of national provisions)
International Law	1. Multilateral and bilateral agreements 2. International regulations of cooperative system • European Outline Convention on Transfrontier Cooperation (1980) • Draft European Charter of Regional Self-government (1997) • European Charter of Local Self-government (1985)

Among Hungary's neighbouring countries on the external borders, Ukraine ratified the Convention on 21st September 1993, but Serbia, where local governments are still undeveloped, has not signed that and the absence of European norms can aggravate the situation. This fact can seriously reduce the chances of cross-border cooperation systems. Usually there is a chance to sign bilateral agreements at local level in countries where the lower levels have a large degree of autonomy.

On the other hand, the Madrid Convention authorises regional and local bodies to sign international treaties only under the supervision of the national level authorities. Consequently, here the state preserves its right of veto, as the exclusive possessor of competence on foreign affairs[12].

It belongs to the democratic principles of integration to acknowledge that local (and regional) self-governments have political power and have the right to participate in the international activities of the State at any time. Local authorities shall be entitled to cooperate with their counterparts in other States in the manner defined by the relevant national legislation, where their interests are concerned. The central government of Serbia, for example, cannot entirely close its eyes to the legal obligation to harmonize domestic law with existing international law. So the Serbian Parliament has to approve the Madrid Outline Convention and the Government to secure conditions for the application of the European Charter of Local Self-Government (1985), particularly the right of local governments to unite at international level. 'Local authorities shall be entitled, under such conditions as may be provided for by the law, to cooperate with their counterparts in other States'[13].

---

12 E. Soós, and Z. Fejes, "Hungarian Experiences of Cross-Border Cooperation", in *Fundamentum*, Vol. 12, No. 5, 2008, p. 125.

13 Council of Europe, *European Charter of Local Self-government*, Strasbourg, European Treaty Series No. 122.15.X, 1985, Art. 10. (3), http://www.conventions.coe.int/Treaty/Commun/ChercheSig.asp?NT=122&CM=8&DF=12/07/2012&CL=ENG [accessed on 1st August 2013].

4. EGTC Establishments on the Ground

4.1. The Failure of Cross-Border Partnership in the Hungarian-Romanian-Serbian Triple-Border Region

The Hungarian-Romanian-Serbian triple-border region can be considered a gradually dynamic region on the edge of the Great Hungarian Plain. Its main centre is Szeged with Mórahalom, a nearby small town, which is active in the renewal of ties with Vojvodina. The formation of the Banat-Triplex-Confinium European Grouping of Territorial Cooperation Limited (BTC EGTC) started at the meeting of 50 mayors from the triple-border region which decided in Mórahalom (Hungary) on 17th June 2009 to establish a Mórahalom-based European Grouping of Territorial Cooperation.

Referring to the EGTC Regulation[14] the creation of an EGTC is the competence of the cooperating parties, and the matter is to be decided by the municipalities. The participative members have to make decisions about the intention to join the grouping of territorial cooperation. Since local governments are decision-making bodies of political authority decisions on cooperation with the local governments of other countries are to be made with a qualified majority within the governing body. On 10th December 2009 upon the initiative of participative members, the Convention[15] and the Statute of BTC EGTC[16] was approved.

The establishment of the EGTC is connected with the process of approval by national authorities – the criteria for approval or rejection by national authorities are specified and the time for examination is limited to three months. The EGTC members must first obtain authorizations from all the states concerned. In Hungary the Approval Authority, the Ministry of Public Administration and Justice, approved the grouping in 2010.

In Romania the Approval Authority is the Ministry of Regional Development and Tourism (MRDT), which issued the approval of Romanian entities to the association at the establishment of an EGTC to be located in Mórahalom (Hungary). Concerning the third part of the prospective EGTC, in 2009 the councils of Vojvodinean municipalities made a decision to join the BTC EGTC. They notified the Serbian

14 European Parliament and Council of the European Union, *Regulation (EC) No. 1082/2006 on the European Grouping of Territorial Cooperation (EGTC)*, *op. cit.*, Art. 3.

15 Banat-Triplex-Confinium European Grouping of Territorial Cooperation Limited, *Convention*, Mórahalom, 10th December 2009.

16 Banat-Triplex-Confinium European Grouping of Territorial Cooperation Limited, *Statute*, 10th December 2009, amended by the Decision 1/2010 No. XII.20, Mórahalom, 20th December 2010.

government about their intention to participate in the grouping but the petition of local authorities was refused first by the Ministry of State Administration and Local Government in September 2009, then by the Ministry of Human and Minority Rights, State Administration and Local Government in May 2011. In the explanation of the Ministry, the main reason of refusal was that Serbian legal system cannot allow the municipalities to participate in such type of grouping. In the official response, the Ministry argues that the Republic of Serbia has no bilateral agreement with the EU member states and there is no applicable national law on EGTC allowing the participation of the Voivodinean municipalities.

Taking into consideration the government resolution the Voivodinean municipalities decided not to slow down the establishment process. Ad interim, until the elimination of legal difficulties, while the Serbian government approves their full member status in the EGTC, they will take the position of an observer. From Serbia eight municipalities from Bačka and Banat joined the EGTC as observers. The Cooperation Agreement between the BTC EGTC and the eight Voivodinean municipalities was signed in Mórahalom on 10th December 2009[17].

The full members of the EGTC obtained the approval of their own states, and the Banat-Triplex-Confinium EGTC was registered in Budapest on 5th of January 2011 with the affirmation of the Budapest Metropolitan Court.

The aim of cooperation on the triple border is to underpin economic and social cohesion with territorial cooperation. The EGTC's main aim is the implementation of territorial cooperation programmes or projects with or without the financial contribution of the Community. The main task of EGTC is to dynamise border areas which were marginalized in the last few decades. So particularly important are the BTC EGTC labour development strategy in the fields of agricultural innovation, the exploitation of renewable energy sources, the infrastructure, the enterprise restructuring, the education and training, and achievements of particular applications of EU funds[18].

How to remove obstacles from the establishment of an EGTC with Serbian full members? The primary concept was the setting up of the triple-border cooperation in the framework of an EGTC, but the dismissal of Serbian central authorities significantly changed the nature of cooperation.

17 Cooperation Agreement between the Banat-Triplex-Confinium European Grouping of Territorial Cooperation Limited and the local governments of Ada, Begaszentgyörgy, Csóka, Kikinda, Magyarcsernye. Magyarkanizsa, Törökkanizsa, Zenta. Mórahalom, 10th December 2009.

18 Banat-Triplex-Confinium European Grouping of Territorial Cooperation Limited, *Convention*, *op. cit.*, p. 2.

However, the EU does facilitate the participation of local and regional authorities from third countries in the European groupings of territorial cooperation. The EGTC Regulation does not explicitly exclude the possibility of the involvement of entities from a third country, or the participation of their regional and local bodies in an EGTC formed in accordance with this regulation. In the examined case the implementation difficulties made the participation for the Vojvodinean municipalities from Serbia impossible. Its main reason stems from a perception of incompatibility of the differing statuses of the neighbouring states. The EGTC Regulation cannot be applied directly in non-EU member states. On 1st March 2012, the European Council agreed to grant Serbia the status of candidate country. Reassuring for the BTC EGTC and the future EGTCs on the Hungarian-Serbian border that the Council's progressive resolution may eventuate the adoption and implementation of the acquis communautaire, among others the 1082/2006/EC Regulation, in the internal legal system as early as during the accession negotiations[19].

Signing agreements with other municipalities beyond the borders is to be the competence of municipalities. The legal background of the international relations of local governments – considering the principles of the Charter of Local Self Government – is guaranteed by the Constitution and the Act on Local Governments in Serbia. 'Autonomous provinces and local self-government units shall cooperate with the corresponding territorial communities and local self-government units from other countries, within the foreign policy of the Republic of Serbia, observing the territorial integrity and legal system of the Republic of Serbia'[20].

Even though the cross-border cooperation for Serbian municipalities is formally guaranteed by the constitutional and legislative frameworks, without the political endorsement of the establishment of the Banat-Triplex-Confinium EGTC and other prospective EGTCs by the central governmental authorities the legal background remains a required but insufficient term.

---

19 European Commission, *Staff working paper. Analytical report. Communication from the Commission to the European Parliament and the Council. Commission Opinion on Serbia's application for membership of the European Union*, COM (2011) 668, pp. 3-5.

20 *Constitution of the Republic of Serbia*, Cooperation of Autonomous Provinces and Local Self-government Units, Article 18. *Survey*. Republic of Serbia. A record of facts and information. http://www.pregled-rs.rs/products2.php?id=1&lang=en [accessed on 29th August 2013], see also *Zakon o lokalnoj samoupravi*, VI Saradnja i udruzivanje jednica lokalne samouprave, Član 88. *Službeni glasnik*, Republike Srbije [Official Journal of the Republic of Serbia], 129/2007 http://www.mrrls.gov.rs/sites/default/files/attachment/zakon_o_lokalnoj_samoupravi.pdf [accessed on 29th August 2013].

Cross-border cooperation among local communities of neighbouring countries is a real indicator of stability of relations and normalisation of relations between the two countries[21]. The failure of the full membership of the Voivodinean municipalities raises a question of political nature. In order to overcome the obstacles hindering cross-border cooperation there is a demand to have an international agreement between the two states, further bilateral agreement between the EU member state Hungary and non-EU Serbia. The lack of bilateral agreements between the two countries restricts the cooperation initiatives and development in the Hungarian-Serbian border area.

Among the few advantages of the observer status for the Vojvodinean municipalities are that they do not have to search for partners for IPA projects any more since they can find partners among the EGTC partners to different IPA programmes. Besides, there is also a hope among partners that the future cross-border cooperation programmes may facilitate the establishment of good and close relations with the Serbian government and herewith built good neighbourly relations.

4.2. Challenges of the Establishment of EGTCs in the Hungarian-Romanian-Slovakian-Ukrainian Border Area

By 1st August 2007 (EGTC Regulation 1082/2006 Art. 18), member states should have implemented their national provisions in order to ensure the effective application of the EGTC. In early 2007, there was an attempt to establish an EGTC in the bordering areas of Hungary, Romania, Slovakia, and Ukraine. The formalised Ung-Tisza-Túr EGTC (UTT EGTC) started functioning in 2007 before the adoption of national EGTC provisions of the involved member countries. This was a unique quadrilateral initiative on the external EU border with the participation of Hungarian, Romanian, Slovakian and Ukrainian partners. The inauguration meeting was held on 27th January 2007. The seat of the UTT EGTC is located in Turistvándi, Hungary. After the unanimous approval of the proposed Convention and Statute by the members they had to submit them to the notification authority. In the examined case some obstacles were encountered in the approval process of the EGTC. The main point of discussion was that after the Romanian municipalities had sent the Romanian central authorities the proposed Convention and Statute, referred to as Articles 8 and 9 of the Regulation, the Ministry of Regional Development and Tourism in Romania refused the petition. They found fault with the Statute

21 J. Jelinčić, and S. Durovič (eds.), *Cross-Border Cooperation as Serbia's Development Instrument*, Belgrade, Centre for Applied European Studies, Fund for an Open Society, 2009, p. 8.

of the UTT EGTC, underlining that the statute contradicts Romanian national laws.

The EGTC made arrangements to correct the questioned rules of the Statute and sent it again to the central authorities. Since then, the Romanian authorities have been keeping back the documents thus continuing to slow down the approval process.

In the meantime another cooperation project was endorsed in the Hungarian-Slovakian border area, the Ung-Tisza-Túr-Sajó (Hernád-Bódva-Szinva) EGTC with Limited Liability (UTTS EGTC). The territory of this EGTC covers Prešov and Košice (from Slovakia), Borsod-Abúj-Zemplén, Szabolcs-Szatmár-Bereg and Hajdú-Bihar counties from Hungary, Satu Mare and Arad counties from Romania and the Subcarpathian territories from Ukraine. The grouping seat is Kántorjánosi in Hungary, the seats of the EGTC operational office are in Janik (Slovakia), Turulung (Romania) and Kaszon (Ukraine). The draft convention and statute have been existing since 30th November 2007[22].

With the same name (UTTS EGTC) the new EGTC was set up on 9th December 2008 in Kántorjánosi, Hungary. The EGTC's participative members are Borsod-Abúj-Zemplén, Szabolcs-Szatmár-Bereg and Hajdú-Bihar counties from Hungary and Prešov and Košice counties from Slovakia. The participation of Romanian and Ukrainian partners has been postponed. The Hungarian-Slovakian cooperation was registered on 5th January 2009 in the Budapest Metropolitan Court.

By the Convention, the cooperation between Hungary and Slovakia shall act for the purpose of implementing territorial cooperation programmes or projects co-financed by the European Union, notably under the structural funds through ERDF, ESF and the Cohesion Fund. Within the framework of the EGTC the main objectives are to give assistance to harmonious development in various fields like renewable energy, ecological rehabilitation, enterprises, common environmental management and protection[23].

Summing up the ongoing process in the north-eastern border area of Hungary it can be claimed that the establishment and operation of a European grouping on the internal border of the EU do not face any special difficulties. But the initiative of municipalities in the quadrilateral border area has failed because practical and to some extent political difficulties have arisen in the application of the EGTC Regulation.

---

22 METIS GmbH, *The European Grouping of Territorial Cooperation. State of Play and Prospects*, Brussels, European Union, 2009, pp. 75-78.

23 Ung-Tisza-Túr-Sajó (Hernád-Bódva-Szinva) EGTC with Limited Liability, *Statute*, Kántorjánosi (Hungary), 9th December 2008, p. 4.

Among the main points of discussion was the liability of the EGTC. In Hungary local government budgetary organisations have limited liability. The local government's partnership with a legal entity and a local government' budgetary organisation is not allowed to exceed the extent of its material contribution[24]. But the Romanian legislation allows the EGTC members to be jointly and unlimitedly liable for EGTC's debts, each member's share being fixed in proportion to its contribution[25].

The Romanian national legislation allows also an exception concerning the participation of third countries which is generally not mentioned in the national provisions. In compliance with Art. 2 (4) pursuant to the provisions of the EGTC Regulation, entities from third countries shall have the right to join the EGTC, providing that their national legislation allows so[26].

As far as the good neighbourly relations are concerned, this is the case when the basic treaties with these countries highlight the intension of friendly cooperation. There is a Treaty between the Republic of Hungary and Ukraine on the Foundations of Good Neighbourly Relations and Cooperation (1991) and one with Romania – the Treaty on Understanding, Cooperation and Good-Neighbourly Relations (1996). These treaties incorporate soft law provisions, reflecting the strong influence of political factors in the adaptation of agreements.

To sum up, it is quite evident that there is still a long way to go to have quadrilateral cooperation along the external borders of Hungary. Participation is still not justified for reasons of public interest or of public policy of member states. The main actors in the promotion of cross-border cooperation are the local governments which recognised that the regions on the border had identical economic and socio-cultural roots and perspectives. They assume that the low level of cooperation in the border region can be raised jointly by means of implementing cross-border projects, therefore they opted to establish an EGTC. But the domestic situation of the neighbouring countries also affected the establishment of the grouping, as at political level it was not welcomed in the neighbourhood of Hungary. Because of the large Hungarian population, there is a great activity on the Romanian side which manifests

---

24 Hungarian Parliament, *Act XCIX. of 2007 on the European Grouping of Territorial Cooperation, op. cit.*, Art. 2 (3).

25 Romanian Government Ordinances, *Emergency Ordinance Regarding the European Grouping of Territorial Cooperation*, Official Journal of Romania, Part 1, No. 769, 13, XI, 2007. Art. 14 (3).

26 European Parliament and Council of the European Union, *Regulation (EC) No. 1082/2006 on the European Grouping of Territorial Cooperation (EGTC), op. cit. Preamble (16).*

itself in more diverse and more intense economic relations and people to people actions[27]. The possible explanation for the resistance of Romanian authorities might be that they consider their sovereignty "threatened" by cross-border cooperation with a strong ethnic component[28]. The negative attitude makes it more difficult to cooperate across the borders in areas with a large number of Hungarian minority populations. Progress in relations could be achieved by signing bilateral intergovernmental agreement between Hungary and Romania on cross-border cooperation, based on the Madrid Outline Convention. (A similar one already exists with Ukraine (1997) and with Slovakia (2001)).

5. Conclusions

Both prospective EGTCs, the Banat-Triplex-Confinium EGTC in the Hungarian-Romanian-Serbian triple-border region and the quadrilateral Ung-Tisza-Túr EGTC in the Hungarian-Romanian-Slovakian-Ukrainian borderline area are still lacking the 'top-down' support of the neighbouring central governments. After having helped EGTC make its first step, the establishment, the central governments prevent them from growing into a strong and mature structure.

In the examined and proposed EGTCs on the external borders of Hungary it was the political reluctance that slowed down the elaboration process. In the case of BTC EGTC it is Serbia who blocks the establishment process, while in the case of UTT EGTC it is an EU member state, Romania. The delay in the approval process is the consequence of the terms of establishment of the groupings determined by the EU regulations and also by the national policies of the involved countries.

Examining the practice on the external borders of Hungary, the cross-border cooperation activities are primarily based not on the activities of the state but on those of the local or regional authorities of a given state. In a Hungarian context, we cannot speak about such mature and effective legal practice yet as the cooperation systems are very rudimentary. The cross-border cooperation systems are basically the expansions of various economic, socio-cultural and political actors.

The contribution of regional and local levels in cross-border relations is determined by the extent the central government broadens their competence in public law. The relationship between local and

27 L. Dancs, "Határon átnyúló kapcsolatok az Északkelet-Alföldön" [Cross-Border Relations in the North Eastern Plain], in B. Baranyi (ed.), *Északkelet Magyarország és a határok [North Eastern Hungary and the Borders]*, Debrecen, Kézirat MTA RKK Alföldi Tudományos Intézete Debreceni Csoport, 1999, pp. 329-370.

28 F. Palermo, "Senior Legal Adviser to the OSCE. High Commissioner on National Minorities", *op. cit.*, p. 13.

regional democracy and cross-border cooperation should be seen as complementary notions: effective cross-border cooperation needs strong local self-government and effective local governance.

As far as their legal status is regarded, the EGTCs in which Hungarian municipalities participate have cross-border legal entity, therefore they can sign regional agreements with international public legal content. Due to the different laws and degrees of centralisation in Hungary's neighbouring states, it is necessary to provide the participants with the legal background of cooperation systems, the scope of competences, and the right to sign contracts. Nevertheless, the cross-border cooperation systems have a natural effect on state sovereignty since certain state spheres of authority include the responsibility for the EGTC setups.

The authorities of the central governments are still convinced that each type of international relations is the monopoly of the bodies of the central government, and no cross-border cooperation on the part of local or regional authorities should be allowed. In consequence, some governments took steps against the cooperation initiatives of their own local and regional authorities. The Serbian and Romanian governments acknowledge the significance of cooperation but try to keep them under state control, favouring the initiatives generated by the central state authorities. The excessively centralized nature of the central power, the nationalist attitude of the governmental leadership facilitate the isolation of border areas.

However, there are several cases where the members, or prospective members, would see advantages in involving members from a third country or even setting up an EGTC between regions or local authorities of two member states and a third country. It is particularly true in those cases where a significant minority lives in the border area and they want to establish close links with their kin-state. The EGTC tool allows the intervention across borders and the motivation to set up an EGTC often depends on the previous history of the countries involved. Obviously, this has a deep impact on the national minorities of a country. Cross-border cooperation "below the nation state" resuscitates regional and local identities in the border areas.

Consequently, the building-up and institutionalisation of cooperation systems are largely influenced by bilateral agreements signed by central governments. That is why it is very important to adopt bilateral and multilateral agreements with neighbouring countries, to ensure the protection of persons belonging to the national minorities concerned.

Third countries, such as Ukraine or Serbia, can be involved in an EGTC-like cooperation system initiated by two EU member states. For third country partners interested in an EGTC to be involved in it is essential to pass a national legal regulation that allows such a form of cooperation. Cooperation with third countries within the framework of the EGTC is possible if it is allowed by the national legal regulation of the given EU member state. The EGTC left the question open, the decision is left to member states.

For the creation and effective operation of cross-border cooperation systems, it is necessary to create the legal frameworks within which these cooperation systems could achieve their objectives. Counterbalancing the dominance of central government, in cooperation systems along the borders, the local and regional authorities come to the fore.

In order to have more EGTCs in the border areas Hungarian national and subnational diplomacy should keep the issue on the political agenda: the expectation towards the Serbian government is to sign the most important international documents, the Madrid Convention and the bilateral agreement with Hungary, and promote more active relations with the Romanian central government to abolish the prejudice and make them more cooperative towards cross-border 'bottom-up' events.

EUROCLIO – Published Books

N° 81 – *The European Community and the World. A Historical Perspective.* Giuliana Laschi (ed.), 2014.

N° 80 – *Christian Democrat Internationalism. Its Action in Europe and Worldwide from Post World War II until the 1990s. Volume II: The Development (1945-1979). The Role of Parties, Movements, People.* Jean-Dominique Durand (ed.), 2013.

N° 79 – *Christian Democrat Internationalism. Its Action in Europe and Worldwide from Post World War II until the 1990s. Volume I: The Origins.* Jean-Dominique Durand (ed.), 2013.

N° 78 – *Investment Banking History. National and Comparative Issues (19th-21st centuries).* Hubert Bonin & Carlo Brambilla (eds.), 2013.

N° 77 – *Citizenship and Solidarity in the European Union. From the Charter of Fundamental Rights to the Crisis, the State of the Art.* Mariana Canotilho, Pedro Madeira Froufe & Alessandra Silveira (eds.), 2013.

N° 76– *The OSCE: Soft Security for a Hard World. Competing Theories for Understanding the OSCE.* Roberto Dominguez (ed.), 2013.

N° 75 – *Euroscepticism, Europhobia and Eurocriticism. The Radical Parties of the Right and Left vis-à-vis the European Union.* Cesáreo Rodríguez-Aguilera de Prat, 2013.

N° 74 – *Fermer les mines en construisant l'Europe. Une histoire sociale de l'intégration européenne.* Nicolas Verschueren, 2012

N° 73 – *Reflets de la construction européenne. Réflexions, références et refus du débat sur l'Europe.* Bertrand Vayssière (dir.), 2012.

N° 72 – *The International Relations of Local Authorities. From Institutional Twinning to the Committee of the Regions: Fifty Years of European Integration History.* Fabio Zucca, 2012.

N° 71 – *Histoire du Conseil de l'Europe (1949-2009).* Birte Wassenberg, 2012.

N° 70 – *The EU, Security and Transatlantic Relations.* Finn Laursen (ed.), 2012.

N° 69 – *Européanisation au XXe siècle. Un regard historique / Europeanisation in the 20th Century. The Historical Lens.* Matthieu Osmont, Émilia Robin-Hivert, Katja Seidel, Mark Spoerer & Christian Wenkel (dir./eds.), 2012.

N° 68 – *Consensus and European Integration. An Historical Perspective/ Consensus et intégration européenne. Une perspective historique.* Daniela Preda & Daniele Pasquinucci (eds./dir.), 2012.

N° 67 – *Générations de fédéralistes européens depuis le XIXe siècle. Individus, groupes, espaces et réseaux*. Geneviève Duchenne et Michel Dumoulin (dir.), 2012.

N° 66 – *Raymond Aron, penseur de l'Europe et de la nation*. Giulio De Ligio (dir.), 2012.

N° 65 – *Europe Twenty Years after the End of the Cold War. The New Europe, New Europes? / L'Europe vingt ans après la fin de la guerre froide. Nouvelle Europe, nouvelles Europes?* Bruno Arcidiacono, Katrin Milzow, Axel Marion & Pierre-étienne Bourneuf (eds./dir.), 2012.

N° 64 – *Aux origines de la diplomatie européenne. Les Neuf et la Coopération politique européenne de 1973 à 1980*. Maria Gainar, 2012.

N° 63 – *Europe and the Middle East. The hour of the East?* Birte Wassenberg & Giovanni Faleg (eds.), 2012.

N° 62 – *Quelle architecture pour un code européen de droit international privé ?* Marc Fallon, Paul Lagarde, Sylvaine Poillot-Peruzzetto (dir.), 2011.

N° 61 – *La construction d'un espace euro-méditerranéen. Genèses, mythes et perspectives*. Houda Ben Hamouda et Mathieu Bouchard (dir.), 2012.

N° 60 – *Innovations et transferts de technologie en Europe du Nord-Ouest aux XIXe et XXe siècles*. Jean-François Eck et Pierre Tilly (dir.), 2011.

N° 59 – *Novel Outlooks on the Marshall Plan. American Aid and European Re-Industrialization*. Fransceca Fauri & Paolo Tedeschi (eds.), 2011.

N° 58 – *Europe in the International Arena during the 1970s. Entering a Different World / L'Europe sur la scène internationale dans les années 1970. à la découverte d'un nouveau monde*. Antonio Varsori & Guia Migani (eds./dir.), 2011.

N° 57 – *Cultures nationales et identité communautaire. Un défi pour l'Europe ? / National Cultures and Common Identity. A Challenge for Europe?* Marloes Beers & Jenny Raflik (dir./eds.), 2010.

N° 56 – *Les trajectoires de l'innovation technologique et la construction européenne. Des voies de structuration durable ? / Trends in Technological Innovation and the European Construction. The Emerging of Enduring Dynamics?* Christophe Bouneau, David Burigana & Antonio Varsori (dir/eds.), 2010.

N° 55 – *L'OTAN et l'Europe. Quels liens pour la sécurité et la défense européenne ?* Birte Wassenberg, Giovanni Faleg et Martin W. Mlodecki (dir.), 2010.

N° 54 – *The Road Europe Travelled Along*. Daniele Pasquinucci and Daniela Preda (eds.), 2010.

N° 53 – *Vivre et construire l'Europe à l'échelle territoriale de 1945 à nos jours*. Yves Denéchère et Marie-Bénédicte Vincent (dir.), 2010.

Euroclio European Network

together with the SEGEI Network

Coordination : Chaire Jean Monnet d'histoire
de l'Europe contemporaine (Cehec)
Collège Erasme, 1, place Blaise-Pascal, B-1348 Louvain-la-Neuve

Germany
Jürgen Elvert
Wilfried Loth

Belgium
Julie Cailleau
Jocelyne Collonval
Yves Conrad
Gaëlle Courtois
Pascal Deloge
Geneviève Duchenne
Vincent Dujardin
Michel Dumoulin
Roch Hannecart
Pierre-Yves Plasman
Béatrice Roeh
Corine Schröder
Caroline Suzor
Pierre Tilly
Arthe Van Laer
Jérôme Wilson
Natacha Wittorski

Spain
Enrique Moradiellos
Mercedes Samaniego Boneu

France
Françoise Berger
Marie-Thérèse Bitsch
Gérard Bossuat
Éric Bussière
Jean-François Eck
Catherine Horel
Philippe Mioche
Marine Moguen-Toursel
Sylvain Schirmann
Matthieu Trouvé
Laurent Warlouzet
Emilie Willaert

Hungary
Gergely Fejérdy

Italy
David Burigana
Elena Calandri
Eleonora Guasconi
Luciano Segreto
Antonio Varsori

Luxemburg
Charles Barthel
Etienne Deschamps
Jean-Marie Kreins
René Leboutte
Robert Philippart
Corine Schröder
Gilbert Trausch

The Netherlands
Anjo Harryvan
Jan W. Brouwer
Jan van der Herst

Poland
Józef Laptos
Zdzisiaw Mach

Switzerland
Antoine Fleury
Lubor Jilek